Principles of Fire Engineering

~

Science, Safety and Solutions

By
Duncan Winsbury

MAPLE
PUBLISHERS

Principles of Fire Engineering Science, Safety and Solutions

Author: Duncan Winsbury

Copyright © 2025 Duncan Winsbury

The author asserts the moral right to be identified as the author of this work.

The right of Duncan Winsbury to be identified as author of this work has been asserted by the author in accordance with section 77 and 78 of the Copyright, Designs and Patents Act 1988.

First Published in 2025

ISBN 978-1-83538-643-9 (Paperback)
     978-1-83538-644-6 (Hardback)
     978-1-83538-645-3 (E-Book)

Cover Design and Book Layout by:
    White Magic Studios
    www.whitemagicstudios.co.uk

Published by:
    Maple Publishers
    Fairbourne Drive, Atterbury,
    Milton Keynes,
    MK10 9RG, UK
    www.maplepublishers.com

A CIP catalogue record for this title is available from the British Library.

All rights reserved. No part of this book may be reproduced or translated in any form or by any means, electronic or mechanical, including photocopying, recording or by any information storage and retrieval system without written permission from the author.

While every effort has been made to ensure accuracy of the information in this book, the author does not assume any responsibility for errors or omissions, accuracy, usefulness, timeliness of content or for completeness. The author disclaims all responsibility and liability to any person for any action or inaction based upon information in this book.'

Acknowledgements

I would like to thank the following, but for whose sharp eyes and assistance, producing this book would have been even more difficult and time consuming:

Aidan Winsbury: *I am proud to say he is my son and a fantastic Mechanical Engineer; he works within the Mechanical & Electrical industry in the UK. Thank you, Aidan, for reviewing this book, (Daniel and Rebecca, I am equally proud of what you both have achieved in life).*

Nick Yates: *Nick is a fire engineer with a sharp eye and a scientific brain. Thank you, Nick, for throwing light on the wider implications of Hydrofluoric Acid issues.*

Colleen, *my wife, for supporting me through the long hours involved in the production of this book and through life as well.*

Foreword

Principles of Fire Engineering - Science, Safety, and Solutions is a comprehensive guide that encapsulates the essential elements of fire engineering. This book is intended to serve as an invaluable resource for fire safety professionals, engineers, building surveyors, architects, policymakers, and students seeking to understand the intricacies of fire science, fire safety design, and regulatory compliance.

Fire engineering is a discipline that has evolved significantly over the centuries, blending scientific principles, engineering innovations, and safety regulations to protect lives and property. As modern structures become increasingly complex, urban populations expand, and climate change introduces new challenges to fire risk management, the need for fire safety has become even more critical.

Through the chapters, readers will explore the fundamental principles of fire dynamics, the movement of smoke, the role of fire protection systems, and the latest advancements in technology that are shaping the future of fire safety. This book incorporates case studies that offer invaluable lessons from historical fire incidents, providing insights into how regulations and fire engineering practices have evolved to mitigate risks. The book also highlights the psychological and behavioural aspects of fire evacuation, ensuring a holistic understanding of fire safety, beyond technical design and regulations.

In a rapidly evolving world, where urbanization and technological advancements bring both opportunities and challenges, fire engineers play a vital role in safeguarding communities. By combining rigorous scientific research with practical applications, this book equips its readers with the knowledge and skills to contribute to a safer built environment.

It is my sincere hope that this book will serve both as an educational tool and as a reference guide for professionals and students alike. As fire safety continues to evolve, let us remain committed to innovation, learning, and collaboration in our pursuit of minimization of fire hazards and saving lives.

Duncan Winsbury

Principles of Fire Engineering
Science, Safety and Solutions

Introduction

Fire is one of the most fundamental, yet complex phenomena encountered in the built environment. It has the power to destroy structures, endanger lives, and cause significant economic losses. However, through scientific understanding, engineering innovation, and well-planned safety measures, fire risks can be mitigated and solutions developed, to ensure the safety of people and property.

Principles of Fire Engineering – Science, Safety, and Solutions is a comprehensive guide that explores the science behind fire behaviour, the safety protocols essential for fire protection, and the engineering solutions that contribute to effective fire prevention and response.

This book is designed for fire safety professionals, engineers, building surveyors, regulatory authorities, and students who seek a deeper understanding of fire engineering. It brings together the latest research, technological advancements, and regulatory requirements to present a holistic approach to fire safety. By bridging the gap between theoretical knowledge and practical application, this book aims to enhance the reader's ability to analyse fire risks, design effective fire safety systems, and implement best practices in fire prevention and control.

A core aspect of fire engineering is the study of fire dynamics, which involves understanding how fire ignites, spreads, and interacts with various materials amidst differing environmental conditions. This book delves into fire science, covering key concepts such as combustion processes, heat transfer mechanisms, smoke movement, and the role of ventilation. Understanding these principles is crucial for predicting fire behaviour and designing efficient fire protection strategies.

In addition to fire science, safety is a central theme of this book. Fire safety measures encompass building design, evacuation planning, detection and suppression systems, and the role of human behaviour in emergency situations. The integration of these elements into building regulations and industry standards ensures that structures are designed and maintained to minimize fire hazards and enhance occupant safety.

Engineering solutions play a pivotal role in mitigating fire risks, from passive fire protection methods such as fire-resistant materials and compartmentation to active systems like sprinklers, alarms, and automated suppression technologies. This book explores advancements in fire engineering, including the use of computational modelling, performance-based design approaches, and smart fire detection systems. These innovations are transforming fire safety, allowing engineers to create more resilient and adaptive fire protection strategies.

Furthermore, the book highlights the importance of interdisciplinary collaboration in fire engineering. Fire safety is not the responsibility of a single profession but requires cooperation between architects, engineers, surveyors, fire service professionals, and policymakers. By fostering a multi-disciplinary approach, fire safety solutions can be more effective, comprehensive, and aligned with societal needs.

Principles of Fire Engineering - Science, Safety, and Solutions aims at being a valuable resource for anyone involved in fire safety and engineering. By equipping readers with essential knowledge and practical insights, this book contributes to the advancement of fire protection practices and the creation of safer built environments. Whether you are a student beginning your journey in fire engineering or a seasoned professional looking to deepen your expertise, this book offers a structured and informative guide to understanding and addressing the challenges posed by fire.

CONTENTS

Introduction   **Page no.**

Chapter 1: The Nature of Fire, its Costs and Consequences 12

FIRE SCIENCE

Chapter 2: Historical Context & Evolution of Fire Safety Engineering.
- Evolution of Fire Safety 26
  - Ancient Civilizations 26
  - Middle Ages 26
  - 18th & 19th Centuries 27
  - Early 20th Century 27
  - Mid to late 20th Century 27
- Contemporary Practices 27
- Significant Historical Fire Incidents and Their Impact 28
  - The Great Fire of London (1666) 28
  - The Triangle Shirtwaist Factory Fire (1911) 28
  - The Cocoanut Grove Fire (1942) 28
  - The MGM Grand Fire (1980) 29
  - The Grenfell Tower Fire (2017) 29

Chapter 3: Fire Engineering Science 30
- Fire Triangle and Fire Tetrahedron 30
- Combustion and Heat Release 30
- Fire Growth and Flame Spread 31
- Heat Transfer Mechanism 32
- Fire Dynamics and Smoke Movement 34
- Structural Fire Resistance 42
- Fire Suppression Mechanism 43

Chapter 4: Fire Dynamics 45
- Stages of Fire Growth 45
  - Ignition (Incipient Stage) 45

- o Growth Stage 46
- o Flashover (Critical Transition Phase)..............................47
- o Fully Developed Stage 48
- o Decay Stage...49
- Fire Development Curve...50
- Heat Transfer Methods 52
    - o Conduction...52
    - o Convection...53
    - o Radiation 54
- Fire Dynamics in Different Environments 55
    - o Compartment vs. Open-Air Fires 55
    - o Influence of Suppression on Fire Dynamics..................56
    - o Interdisciplinary Relevance...56
    - o Lessons Learned 56
- Case Studies in Fire Dynamics..57
    - o Grenfell Tower Fire (2017, UK) 57
    - o Bradford City Stadium Fire (1985, UK)........................58
    - o Station Nightclub Fire (2003, USA).............................59
    - o MGM Grand Fire (1980, USA) 61

Chapter 5: Smoke Movement and Management....................63
- Components of Smoke Management......................................63
- Axisymmetric Plume Management 66
- Balcony Spill Plume Management 69
- Window Plume Management 72
- Communicating Space Smoke Plume Management................76
- Management of Smoke Flow from Smoke Layer....................79
- Management of Airflow to Control Smoke Flow from Plume............83

SAFETY

Chapter 6: Protecting Lives and Property.............................88
- Pro-active Measures to Prevent Fires 88
    - o Design..89
    - o Active Fire Protection Systems......................................90
    - o Education, Training and Emergency Preparedness.......92

- o   Emergency Response Planning 93
- o   Compliance & Regulatory Bodies 95
- o   Fire Safety Codes & Regulations 95
- Post Incident Analysis 97

Chapter 7: Fire Risk Management 98
- Fire Risk Assessment Principles 98
- Role of Fire Risk Assessors and Engineers 98
- Hazard Identification and Risk Evaluation 98
- Fire Protection Strategies 98
- Emerging Challenges in Fire Risk Management 98
- Integration of Fire Engineering and Fire Risk Assessment 101

Chapter 8: Building Evacuation and Human Behaviour 106
- Egress & Life Safety Design Concept 106
- Evacuation Modelling 107
- Evacuation Strategies 109
- Evacuation Analysis 112
- ASET vs. RSET 113
- Evacuation Time 114

Chapter 9: Fire Protection Systems 120
- Active Fire Protection 120
- Passive Fire Protection 122
- Sprinkler Systems 125
- Fire Alarm and Detection Systems 130

Chapter 10: Building Codes and Regulations 140
- Legislation Governing Fire Safety 140
  - o   The Regulatory Reform (Fire Safety) Order 2005 140
  - o   The Fire Safety Act, 2021 143
  - o   The Fire Safety (England) Regulations 2022 146
  - o   Building Regulations 2010 (Part B)-Fire Safety Requirements .. 150
- High-rise Buildings & the Aftermath of Grenfell 151
- Fire Safety Management in Residential & Commercial Buildings 151
- Enforcement & Penalties 152

SOLUTIONS

Chapter 11: Fire Safety Engineering Design 157
- Standards & Regulations 157
- Egress Design & Evacuation Planning Use of Fire Barriers & Compartmentalization 159
- Fire Safety & Modern Building Design 161
- Performance-Based Building Design 162

Chapter 12: Building Materials - Properties and Performance 174
- Fire Resistance of Building Materials 174
- Reaction to Fire Testing 177
- Innovations in Fire-Resistant Materials ....
- Material Selection for Fire Safety ....

Chapter 13: Building Compartmentation 195
- Objectives of Fire Compartmentation 196
- Fire Engineering Principles behind Compartmentation 196
- Fire Compartmentation Strategy 198

Chapter 14: Fire Investigation 200
- Role of Fire Investigators 200
- Types of Fire Investigation Roles 201
- Techniques for Determining Fire Cause 201
- Scene Examination 201
- Legal Implications & Reporting 204
- Fire Investigation Reports 205
- Legal Standards & Best Practices 205

Chapter 15: Emerging Technologies in Fire Engineering 206
- Smart Building Technologies 206
- Fire Modelling Software and Simulations 207
- Future Trends in Fire Safety Solutions 207

Chapter 16: Fire Strategies 209
- Objectives of a Fire Strategy 209
- Components of a Fire Strategy 209
- Fire Strategy Development & Implementation 211
- Regulation 38 of the Building Regulations 2010 212

- Approved Document B of the UK Building Regulations..................216
- BS 9999: Fire Safety In The Design, Management & Use Of Buildings..................217
- BS 9991: Fire Safety In The Design, Management & Use Of Residential Buildings..................219
- BS 7974: Application Of Fire Safety Engineering Principles To The Design Of Buildings..................222

Chapter 17: Conclusion..................225

Appendices

1. Glossary of Fire Engineering Terms..................229
2. Resources for Further Reading..................234
3. Professional Organisations..................235
4. Fire Engineering Symbols and Formulas..................239
5. Example Reports:..................241
   - A. PAS 79 Fire Risk Assessment..................242
   - B. Qualitative Fire Risk Assessment..................246
   - C. Semi Quantitative Fire Risk Assessment..................250
   - D. Quantitative Fire Risk Assessment..................255
   - E. Fire Strategy Approved Document B..................259
   - F. Fire Strategy BS 9999..................263
   - G. Fire Zone Modelling..................267
   - H. Evacuation Modelling..................270
   - I. CFD Modelling..................273
   - J. Fire Investigation..................276

Chapter 1
The Nature of Fire, Its Costs and Consequences

1.1 Introduction

This introductory chapter lays the foundation for understanding fire as a complex, multifaceted, and ever-present force in human society. Far more than a natural hazard, fire is a phenomenon that intersects with nearly every aspect of human existence, from ancient rituals and technological innovation to disaster response, economic planning, and environmental management. To fully grasp the role of fire in the modern world, it is essential to begin with a broad, interdisciplinary perspective. This chapter serves to establish that perspective and to position fire as both a scientific subject and a societal challenge.

At the most fundamental level, the chapter introduces fire from a technical and scientific viewpoint. Fire is the result of combustion, a chemical reaction involving heat, fuel, oxygen, and an ongoing chemical chain reaction. This concept, captured in the fire tetrahedron, underpins nearly all fire behaviour and response strategies. Without a firm understanding of the elements that sustain fire, one cannot begin to assess how fires start, grow, or can be extinguished. Therefore, this chapter offers an entry point into fire science for readers who may come from a variety of professional backgrounds, including engineering, architecture, urban planning, emergency management, and public policy.

Beyond the science, fire's broader consequences and applications are central to this chapter's purpose. Throughout history, fire has played a dual role. It has enabled human civilisation to develop, providing light, warmth, cooked food, forged tools, and industrial progress. Yet it has also exacted a significant toll, often serving as a reminder of the limits of human control. Catastrophic fires, from ancient urban infernos to modern high-rise and wild land fires, illustrate the destructive capacity of an uncontrolled flame. This duality is a core theme of fire engineering: understanding how to harness fire's benefits while reducing its dangers.

In addition, the chapter touches on the far-reaching impacts of fire, which span physical, social, economic, and environmental domains. Fires cause hundreds of thousands of deaths and injuries worldwide each year, and the emotional, cultural, and economic aftermath can be as devastating as the flames themselves. From displacing entire communities to eroding national infrastructure, fire's reach extends well beyond the incident scene. Moreover, fire interacts with other global challenges, such as climate change, urbanisation, energy demands, and material science, making it a subject of growing complexity.

The objective here is not only to define and describe fire, but to underscore its continuing relevance in the 21st century. The modern built environment presents new risks and demands novel solutions. As buildings become taller, more densely occupied, and more technologically complex, fire safety must evolve in tandem. In parallel, climate-driven wildfires have become more frequent and intense, often breaching the urban-wildland interface and threatening entire communities. These scenarios require holistic fire strategies that account for both engineering controls and human behaviour.

This chapter also introduces the necessity of interdisciplinary collaboration in addressing fire risks. Engineers, architects, fire services, regulators, insurers, and community stakeholders each have a role to play. No single discipline holds all the answers, and fire safety cannot be achieved in isolation. As such, the chapter sets a collaborative tone for the rest of the book, which delves into scientific principles, cost-benefit analysis, regulatory frameworks, system design, and emerging technologies. The intent is to foster a systems-thinking mindset in readers, equipping them to see fire not just as a technical problem, but as a multidimensional issue requiring integrated solutions.

Finally, by exploring the nature of fire and its costs and consequences, this chapter seeks to engage both new and experienced readers. For those unfamiliar with fire engineering, it provides a comprehensive yet accessible overview. For those with professional expertise, it serves as a reminder of the foundational context in which specialised knowledge must operate. Whether one is designing safer buildings, crafting legislation, responding to emergencies, or conducting fire research, this chapter offers a shared starting point.

In doing so, it underscores the central message of the book: that understanding the principles of fire engineering begins with a deep appreciation of what fire is, what it does, and what it costs us—not only in terms of money or materials, but in lives, livelihoods, and ecosystems. Only by confronting the full nature of fire can we hope to develop meaningful, sustainable solutions.

1.2 Defining Fire

To understand fire engineering, one must first understand fire itself, not merely as an event, but as a chemical and physical process governed by fundamental laws of nature. Fire is commonly defined as a rapid oxidation reaction that produces heat, light, and various chemical by-products. It is this reaction that gives rise to flames, the visible evidence of energy being released. Yet this simple definition belies the complexity behind the phenomenon. Fire is simultaneously a source of energy, a threat to life and property, a tool for survival, and a force of nature that interacts with its surroundings in intricate ways.

At the heart of fire is combustion, a chemical reaction between a fuel and an oxidising agent, usually oxygen in the air. For combustion to occur, three basic elements must be present: heat, fuel, and oxygen. This is classically illustrated by the fire triangle. However, to sustain fire, a fourth element, an ongoing chemical chain reaction, must also be present. This addition transforms the triangle into the more comprehensive fire tetrahedron, which represents the self-perpetuating nature of flaming combustion.

Each element of the tetrahedron plays a specific role. Fuel refers to any combustible material, whether solid, liquid, or gas. Its state, moisture content, and chemical composition affect the ease with which it burns. Heat is required to raise the fuel to its ignition temperature. This can be introduced through open flame, electrical energy, friction, or even chemical reactions. Oxygen, typically from air, supports the oxidation process. Without sufficient oxygen, combustion may be incomplete, leading to the production of smoke and toxic gases such as carbon monoxide. Finally, the chemical chain reaction refers to the exothermic feedback loop that maintains the fire by continually producing heat, which then sustains further fuel ignition.

The ignition of fire can be categorised as either piloted ignition, where a spark or flame initiates combustion, or autoignition, where the material ignites spontaneously due to high temperatures without an external flame. Different materials have different ignition temperatures, and understanding these thresholds is essential for fire risk assessment.

Fire behaviour is also influenced by the mode of heat transfer: conduction, convection, and radiation. These mechanisms determine how heat spreads from the point of origin to adjacent materials or compartments. Conduction moves heat through solid materials; convection involves the movement of hot gases or liquids, often responsible for transporting flames vertically; and radiation allows heat to travel across open space, enabling the ignition of materials at a distance from the flame source. A firm grasp of these concepts is essential for predicting fire growth and planning effective compartmentation in building design.

Furthermore, the phases of fire development offer a structured way to understand its progression. These include:

- **Incipient Stage** – the earliest phase, marked by heat build-up and invisible off-gassing of combustible vapours.

- **Growth Stage** – where visible flames appear and fire spreads due to increased fuel involvement and ventilation.
- **Fully Developed Stage** – the peak of fire intensity, where all combustible materials in a compartment may be alight.
- **Decay Stage** – where fuel is exhausted or oxygen becomes limited, reducing flame activity but often increasing toxic smoke output.

A critical turning point in the growth phase is flashover, the sudden transition of a room from localised burning to full-room involvement, triggered by the accumulation of heat and combustible gases. Flashover represents one of the most dangerous moments in structural fires, as survivability within the compartment drops to near zero.

It is also important to distinguish smouldering combustion from flaming combustion. Smouldering occurs without a flame, at lower temperatures, and often in porous solids like foam, textiles, or insulation materials. Though slower, smouldering fires can be more insidious, producing large amounts of smoke and potentially transitioning into flaming fires if disturbed or if oxygen is suddenly introduced.

Modern fire science has extended beyond these foundational principles to include advanced modelling of fire dynamics, accounting for variables such as ventilation, building geometry, fire load, and material properties. Computational tools allow fire engineers to simulate fire scenarios, assess risk, and design mitigation systems with precision. Nevertheless, all these sophisticated analyses rest upon a thorough understanding of the basic definition and properties of fire.

From an engineering perspective, it is essential not only to understand how fire starts and spreads, but also how it interacts with human-occupied environments. The combustibility of materials, the rate of heat release, smoke production, and flame spread characteristics must all be considered when designing safe buildings and infrastructure. The choice of materials, layout of spaces, and integration of suppression systems all relate back to the core scientific understanding of fire.

In summary, defining fire is more than describing flames, it is about understanding an energetic chemical process that has both predictable and unpredictable aspects. It is about recognising the mechanisms of heat transfer, the conditions for ignition, and the phases of development that determine fire behaviour. With this knowledge, fire can be better anticipated, modelled, controlled, and ultimately, mitigated.

1.3 A Dual-Edged Phenomenon

Fire is one of the earliest and most transformative discoveries in human history. Its harnessing marked a pivotal shift in the development of civilisation, enabling early humans to cook food, stay warm in cold climates, ward off predators, and extend productive hours into the night. Archaeological evidence shows that controlled fire use dates back nearly a million years, suggesting that it was as much a part of human evolution as the development of tools or language. From this foundational point onward, fire has played a dual role, both as a powerful enabler of progress and a constant source of danger and destruction.

On the one hand, fire has been a driver of human advancement. With fire, early humans gained access to previously inedible foods, improving nutrition and survival rates. The control of fire led to more permanent settlements, where hearths became central gathering points. In later centuries, fire underpinned major societal shifts, from the smelting of metals during the Bronze and Iron Ages to the invention of steam engines in the Industrial Revolution. Entire industries, including transportation, manufacturing, and energy, owe their origins and development to the controlled use of fire.

Fire also holds cultural and symbolic significance. In many mythologies and religious traditions, fire is a sacred element, representing purification, transformation, or divine power. From the eternal flames of ancient temples to Olympic torches and ceremonial bonfires, fire carries meanings that extend far beyond its physical properties. It has been a metaphor for passion, rebirth, knowledge, and even the human spirit itself.

Yet despite its many benefits, fire has always had a destructive side. It is one of the most unforgiving natural forces, capable of reducing entire communities to ashes in minutes. History is littered with examples of catastrophic fires, often exacerbated by poor planning, inadequate construction materials, or delayed responses.

Some of the most infamous urban fires include:

- **The Great Fire of London (1666)**, which destroyed over 13,000 homes and reshaped the city's architecture and fire laws.
- **The Chicago Fire (1871)**, which left over 100,000 people homeless and prompted major reforms in building codes and fire safety.
- **The Triangle Shirtwaist Factory fire (1911)** in New York City, where locked doors and poor evacuation planning led to the deaths of 146 garment workers, most of them women, an event that triggered sweeping changes in workplace safety regulations.

In more recent memory, events such as:

- **The Grenfell Tower fire (2017)** in London, which claimed 72 lives and exposed systemic failures in building cladding safety, enforcement of fire regulations, and social housing policy.

- **The Camp Fire (2018)** in California, the deadliest and most destructive wildfire in the state's history, which destroyed the town of Paradise and led to renewed scrutiny of energy infrastructure, land management, and climate policy.

These examples show that fire, when unmanaged or misunderstood, can expose vulnerabilities in our systems, whether they be physical, social, or institutional. They also illustrate that the consequences of fire often extend far beyond the flames themselves, uncovering issues related to equity, governance, and resilience.

Today, the dual nature of fire is more evident than ever. On the one hand, controlled fire continues to serve essential functions in industry, agriculture, and energy production. Prescribed burns are used to maintain healthy ecosystems, reduce fuel loads, and prevent larger wildfires. In metallurgy, combustion enables the creation of advanced alloys and tools. In power generation, fire underlies thermal systems that provide electricity and heating to millions.

On the other hand, the risks posed by uncontrolled fire are increasing, not diminishing. The modern built environment, characterised by high-rise buildings, complex electrical systems, and synthetic furnishings, can allow fires to escalate rapidly and release highly toxic smoke. Meanwhile, climate change has intensified the frequency and severity of wildfires across the globe, with longer fire seasons, higher fuel loads, and drier conditions. As human settlements expand into forested and rural areas, the wildland-urban interface (WUI) has emerged as a critical zone of vulnerability.

Moreover, certain types of fire incidents are becoming more difficult to manage. Fires involving electric vehicles, lithium-ion batteries, or renewable energy storage systems introduce new hazards, including thermal runaway reactions, re-ignition risks, and toxic gas emissions. These technologies, while necessary for sustainability, require new approaches in fire detection, suppression, and emergency response.

The dual-edged nature of fire means that fire engineering cannot simply be reactive, it must be proactive, predictive, and adaptive. Professionals must recognise that fire is not an anomaly to be eliminated, but a natural phenomenon to be understood and incorporated into the design and operation of safe systems. This understanding is particularly important as we confront new global challenges, from urban density and climate volatility to emerging technologies and changing demographics.

In summary, fire is both a symbol of human potential and a stark reminder of our limits. Its history is one of progress and tragedy, innovation and loss. To navigate the complexities of fire in the modern world, we must embrace its dual nature—learning from the past while preparing for the future. This balance is at the heart of fire engineering, and it begins with an honest acknowledgment of fire's power to create and to destroy.

1.4 The Scope of Fire Impacts

Fire, in its uncontrolled form, is one of the most destructive forces known to humanity. Its consequences are multifaceted, touching nearly every domain of life, from human health and well-being to built environments, ecological systems, and economic structures. Understanding the full scope of fire's impacts is essential not only for fire professionals but for decision-makers across all levels of society. This section explores the primary domains in which fire exerts influence, highlighting the scale, diversity, and interconnectivity of its effects.

Human Life and Health

Perhaps the most tragic and visible consequence of fire is the loss of human life. Every year, hundreds of thousands of people are killed or injured by fire. The World Health Organization estimates that fire-related burns alone account for over 180,000 deaths annually, the majority occurring in low- and middle-income countries where fire safety infrastructure is limited or absent. These figures exclude deaths caused indirectly by fire, such as those resulting from structural collapse, smoke inhalation, or post-incident trauma.

Beyond fatalities, fire inflicts widespread physical harm. Survivors often suffer severe burns, permanent disfigurement, respiratory damage from inhaled smoke, or psychological trauma. Burns can require extensive surgeries, skin grafts, and long-term rehabilitation. Inhalation of toxic gases, such as carbon monoxide, hydrogen cyanide, and volatile organic compounds, can cause brain damage, organ failure, and long-term disability.

Particularly vulnerable populations include the elderly, children, individuals with disabilities, and those in institutional or overcrowded settings. Fires in hospitals, care homes, and schools have proven particularly devastating due to challenges in evacuation and the high dependency of occupants.

Property and Infrastructure

Structural fires cause billions of pounds in damage each year. Residential fires destroy homes and personal belongings, while commercial fires disrupt business continuity and economic activity. Industrial fires, especially in chemical plants, warehouses, and energy facilities, can trigger cascading effects on supply chains, transportation networks, and regional economies.

Infrastructure is especially susceptible to fire's indirect impacts. Fires often lead to the failure of electrical systems, gas mains, telecommunications, and water supply lines. These disruptions can extend far beyond the immediate site of the fire, affecting entire districts or cities. For example, a transformer fire or underground utility vault fire can bring down power to hospitals, transit systems, and emergency services.

The materials used in modern construction, especially those involving synthetic polymers, foamed plastics, and composite claddings, often burn with high heat release rates and generate dense, toxic smoke. The Grenfell Tower fire in 2017 tragically highlighted how inappropriate cladding and insufficient fire compartmentation can transform a single-point ignition into a full-scale building failure.

Environmental Effects

Fires also leave a lasting mark on the natural environment. Wildfires, whether sparked by lightning, human negligence, or intentional burning, consume millions of hectares of forest and grassland annually. In doing so, they destroy critical habitats, reduce biodiversity, and alter soil chemistry, often leading to increased erosion and flooding during subsequent rainy seasons.

The air pollution caused by fire is a major public health and climate concern. Wildfires and structural fires release significant quantities of particulate matter (PM2.5 and PM10), carbon dioxide, methane, and black carbon into the atmosphere. These emissions degrade air quality, affect respiratory health, and contribute to global warming.

In aquatic environments, fire debris, ash, and firefighting chemicals often find their way into rivers, lakes, and groundwater systems, impacting water quality and threatening aquatic life. Post-fire landscapes can suffer from long-term hydrological disruption due to changes in vegetation and soil structure.

Furthermore, when hazardous substances are involved, such as in chemical plant fires or industrial accidents, the environmental damage can be extreme and long-lasting, with toxic residues remaining in the air, soil, and water for years.

Economic Burden

The economic cost of fire extends far beyond the visible damage. Globally, fire-related losses are estimated to exceed hundreds of billions of pounds annually. This includes direct costs (e.g., property loss, reconstruction, equipment replacement) and indirect costs (e.g., lost income, increased insurance premiums, supply chain disruptions, regulatory fines, and litigation).

In many cases, fire incidents precipitate long-term financial instability for individuals and small businesses. The uninsured or under-insured often face devastating losses, while corporations may experience reputational harm, legal exposure, and shareholder pressure following major incidents.

Governments also bear a significant economic burden through public firefighting services, emergency medical response, disaster relief programs, environmental remediation, and infrastructure repair. In high-income countries, the investment in fire prevention and mitigation is considerable, but so too is the cost of failure when such systems are ineffective or underfunded.

Insurance data offers a window into fire's economic footprint. Insurers regularly report that fire-related claims constitute some of the most expensive and complex to resolve, particularly in large-scale residential developments, data centres, and critical infrastructure projects. Additionally, the cost of compliance with evolving fire codes and standards adds to the economic weight carried by developers and facility managers.

Social and Psychological Impacts

The emotional and psychological toll of fire is often underestimated. Survivors may experience post-traumatic stress disorder (PTSD), anxiety, depression, and a profound sense of loss—especially when homes, heirlooms, and places of personal significance are destroyed. Fires that displace families or communities can create long-term instability, increasing the risk of homelessness, unemployment, and social fragmentation.

In events involving multiple fatalities or mass displacement, fire can leave an imprint on the collective consciousness of a society. Media coverage, public inquiries, and political debates often follow major fire incidents, triggering changes in public policy and civic expectations. Fire disasters can erode trust in institutions, particularly when they reveal lapses in regulation, enforcement, or emergency preparedness.

Fire also disproportionately affects marginalised communities, where older housing, poor access to fire services, and limited financial resilience amplify vulnerability. In this way, fire is not just a hazard, it is also a lens through which social inequality can be observed and addressed.

Understanding the scope of fire's impacts is not a purely academic exercise. It is central to developing effective policies, designing safer structures, managing emergencies, and allocating resources efficiently. Whether measured in lives, pounds, hectares, or carbon emissions, the cost of fire is vast and varied. This understanding forms the rationale for why fire engineering, safety science, and risk management must remain a global priority.

1.5 Why Fire Still Matters in the 21st Century

Despite advances in technology, regulation, and safety design, fire remains a significant and persistent threat in the 21st century. While the context and character of fire risks have evolved, the fundamental challenges posed by fire, its unpredictability, speed, destructiveness, and capacity to overwhelm systems, remain. Fire continues to claim lives, destroy critical infrastructure, degrade environments, and destabilise economies. Its relevance today is arguably greater than ever due to the changing conditions of modern life, including rapid urbanisation, climate change, emerging technologies, and socio-economic disparities.

Urbanisation and Density

Globally, more than half of the population now lives in urban areas, a figure projected to rise to nearly 70% by 2050. With this increase comes a greater concentration of people, buildings, vehicles, and electrical systems in compact spaces. High-density residential developments, particularly in megacities, present a unique set of fire risks. In such environments, fires can spread quickly, and evacuation becomes more complex. Limited access for fire appliances, poor infrastructure, and inadequate enforcement of building codes further complicate the picture.

Informal settlements, common across parts of Asia, Africa, and South America, are particularly vulnerable. These communities often consist of makeshift housing built with combustible materials, tightly packed together, and lacking formal fire protection systems or emergency response infrastructure. In such settings, a single kitchen fire can escalate into a major conflagration affecting hundreds of homes.

Even in high-income nations, increasing urban complexity poses fire safety challenges. Mixed-use buildings, underground structures, and high-rise towers introduce vertical evacuation concerns, smoke control requirements, and fire service access issues. As architecture and urban form continue to evolve, fire safety must adapt in lockstep.

Climate Change and Wildfires

One of the most pressing and visible indicators of fire's continued relevance is the global surge in wildfire activity. Longer fire seasons, hotter temperatures, reduced humidity, and changing vegetation patterns—largely attributed to climate change, have resulted in more frequent, intense, and widespread wildfires. These fires have devastated communities, displaced populations, and destroyed critical ecosystems in countries such as Australia, the United States, Canada, Greece, and Portugal.

The wildland-urban interface (WUI), where natural landscapes meet developed areas, has become one of the most hazardous zones in fire management. As more people move into these regions, drawn by affordable housing, scenic views, or proximity to nature, they unknowingly expose themselves to a higher risk of wildfire. Traditional firefighting techniques are often ineffective in these scenarios, as fires can move at incredible speeds and cross natural and artificial barriers with ease.

In addition to immediate destruction, wildfires contribute significantly to greenhouse gas emissions and air pollution. They degrade air quality over vast distances, increase hospital admissions for respiratory conditions, and contribute to premature deaths. As climate models predict more extreme weather events, fire professionals must prepare for a future where wildfire risk is more than a seasonal concern, it is a year-round threat.

Emerging Technologies and New Fire Risks

Technological innovation has introduced both opportunities and challenges for fire safety. While advances in detection, suppression, and materials science have improved prevention and response capabilities, new technologies often carry novel fire hazards.

Electric vehicles (EVs) and lithium-ion batteries, for instance, present thermal runaway risks that are not easily addressed with traditional firefighting methods. Fires involving EVs can reignite hours or days after initial suppression and often require specialist knowledge and equipment. Battery fires emit toxic gases and high temperatures, placing firefighters and the public at increased risk.

Photovoltaic panels, energy storage systems, and other renewable technologies integrated into modern buildings have added further complexity. While these systems contribute to sustainability goals, they introduce electrical and thermal risks, particularly if not properly installed or maintained. The rise of "smart homes" and integrated IoT devices also introduces fire safety concerns related to overloading circuits, software malfunction, and remote access vulnerabilities.

Data centres, which underpin the global digital economy, have emerged as another critical fire risk area. A single fire in such a facility can lead to vast financial losses, disruptions in cloud services, and potential exposure of sensitive information. Fire suppression in these environments must be both effective and non-destructive, often relying on clean agent systems and redundant design.

Global Inequity in Fire Protection

Another reason fire still matters is the stark inequality in fire safety standards and resources across the world. In many developing countries, fire protection infrastructure is limited, building regulations are inconsistently enforced, and fire services are under-resourced. Slum fires, industrial disasters, and mass-casualty events are tragically common in these contexts.

Even within high-income countries, vulnerable populations, such as the elderly, disabled, low-income families, and minority communities, often live in older, more fire-prone buildings. They may also lack access to education, insurance, or services that would mitigate fire risk. As fire is increasingly viewed through the lens of social justice and public health, addressing these disparities becomes a matter not just of policy, but of ethics.

Changing Human Behaviour and Expectations

Modern society's expectations around fire safety have also changed. People expect a high degree of protection in their homes, workplaces, and public spaces. However, this expectation sometimes leads to complacency. Over-reliance on technology, underestimation of personal responsibility, and reduced public awareness about fire risks have all contributed to dangerous situations.

Meanwhile, changes in behaviour, such as the widespread use of portable heaters, candles, home cooking appliances, and charging devices, have introduced new ignition sources into domestic settings. The COVID-19 pandemic, for instance, saw an increase in home fires as people spent more time indoors, sometimes with limited attention to safety.

The Continued Relevance of Fire Engineering

All these factors reinforce the importance of fire engineering as a contemporary discipline. Fire safety is not a solved problem; it is an evolving challenge that must respond to global trends in technology, climate, demography, and urban planning. The need for holistic, performance-based, and systems-oriented fire strategies has never been greater.

Modern fire engineering must not only focus on protecting life and property, but also on sustainability, resilience, and equitable outcomes. It must draw upon data, modelling, and real-world case studies to create adaptive, evidence-based solutions. And it must engage with architects, regulators, emergency services, insurers, and communities to ensure that safety is integrated from design through to operation and eventual decommissioning.

1.6 A Call to Action

Understanding fire is not merely an academic exercise, it is a moral, professional, and societal imperative. Fire, as demonstrated throughout this chapter, is more than a chemical reaction or a hazard to be managed. It is a dynamic force that has shaped human civilisation, and it continues to test the boundaries of our engineering, policy, and emergency response systems. As such, this chapter concludes with a call to action: a challenge to everyone involved in the built environment, safety governance, and public welfare to recognise the enduring risks of fire and to commit to addressing them with clarity, urgency, and innovation.

One of the fundamental takeaways from this chapter is that fire does not respect boundaries, geographic, economic, disciplinary, or technological. Its consequences ripple across sectors and societies. Therefore, fire safety cannot be the domain of fire services alone. While firefighters remain on the frontlines of incident response, the broader responsibility for fire prevention and mitigation lies with a much larger community of stakeholders.

Fire engineers must lead the way in designing buildings and systems that account for modern fire loads, changing material behaviours, and the increasing complexity of infrastructure. They must ensure that fire modelling, suppression systems, detection technologies, and passive design elements are applied with precision and foresight.

Architects and designers must go beyond aesthetics and functionality to consider how their choices influence fire risk. The layout of escape routes, the materials specified in construction, and the integration of active systems all contribute to the safety of occupants in the event of a fire.

Regulators and policymakers must create and enforce codes that reflect the latest knowledge in fire science and the realities of diverse building types and community needs. They must also ensure that these codes are dynamic, capable of evolving in response to emerging risks such as climate-induced wildfires, green energy infrastructure, and new construction technologies.

Developers and property owners have a crucial role to play in prioritising fire safety, even when it conflicts with cost-saving measures or aesthetic preferences. Investment in fire protection should not be seen as an added cost, but as a long-term value proposition, protecting lives, assets, and reputations.

Insurance providers must continue to analyse fire risk through actuarial models while also promoting loss prevention through education, incentives, and partnerships. Their data and influence can drive safety improvements on a broad scale, especially in regions where regulation is weak or inconsistent.

Educators and researchers must prepare the next generation of fire professionals by embedding interdisciplinary thinking into training programmes. Fire engineering education should include not just science and technology, but also human factors, ethics, economics, and environmental awareness.

Communities and individuals must also be engaged. Public education campaigns, community risk reduction programmes, and local resilience planning are essential to ensure that people are aware of fire hazards and equipped to respond appropriately. Fire safety should be accessible, inclusive, and culturally informed.

This chapter has outlined the nature of fire, its costs and consequences, and its continued relevance in the 21st century. It has demonstrated that fire is not a relic of the past, nor is it a purely technical concern. It is an ongoing challenge that intersects with climate change, urbanisation, inequality, innovation, and health. To meet this challenge, a collective response is needed, one that bridges professional disciplines, national borders, and societal sectors.

The remainder of this book will explore how such a response can be realised. Subsequent chapters will examine the scientific foundations of fire behaviour, delve into the mechanisms of fire detection and suppression, analyse building design principles, and explore the societal and economic dimensions of fire risk. Case studies, regulatory frameworks, and future trends will be presented to illustrate what works, what doesn't, and what must come next.

But before turning the page, let this be a moment of reflection: fire is a force we have lived with for millennia. We have used it to build and been undone by it in equal measure. Our challenge today is not to eliminate fire, an impossible task, but to understand it so thoroughly that we can anticipate, contain, and coexist with it safely.

This is the purpose of fire engineering. This is the promise of science, safety, and solutions.

Chapter 2

Historical Context & Evolution of Fire Safety Engineering

This chapter addresses the development of fire safety measures from the times of ancient civilizations till contemporary times. It goes on to list major fire accidents over the years that were milestones in the development of appropriate safety measures and regulations.

Evolution of Fire Safety

1. Ancient Civilizations

Early Awareness: Fire has been both a vital tool and a hazard since prehistoric times. Early humans learned to control fire for warmth, cooking, and protection.

Primitive Measures: Ancient civilizations like the Roman Empire, Egyptians and Greeks recognized the dangers of fire. They developed basic measures, such as building regulations that specified materials and construction techniques, to reduce fire risks in homes, offices and temples. The Roman Empire even developed residential buildings over 18 metre in height and had a fire brigade that responded to fire incidents for the wealthy, against payment.

2. Middle Ages

Urbanization and Fire Risks: As towns grew, so did the risk of large-scale fires. Wooden structures and close quarters made cities particularly vulnerable.

Fire Watchers: In medieval Europe, towns appointed fire watchers to patrol streets, looking for signs of fire. Citizens were often organized into volunteer fire brigades, similar to the Roman Empire. These fire brigades operated on an insurance scheme and would respond if you were a member.

Regulatory Measures: The Great Fire of London in 1666 led to significant changes in building regulations, including the use of brick and stone instead of wood.

3. 18th and 19th Centuries

Professional Fire Departments: The establishment of organised fire brigades marked a shift in fire response. The first paid fire department was founded in Boston (USA) in 1678 and others followed in major cities.

Fire Insurance: The rise of fire insurance companies in the 18th century encouraged property owners to invest in fire prevention measures to reduce premiums.

Fire Codes: By the late 19th century, formal fire codes began to emerge, focusing on construction materials, occupancy limits, and safety practices.

4. Early 20th Century

Advancements in Fire Fighting: The introduction of motorised fire engines and better firefighting equipment improved response times and effectiveness.

National Fire Protection Association (NFPA), USA: Founded in 1896, the NFPA began developing codes and standards to promote fire safety across various sectors.

5. Mid to Late 20th Century

Fire Safety Engineering: The discipline of professional fire safety engineering emerged, focusing on the science of fire dynamics and the design of safer buildings. In the UK, this was primarily after 2005, when the Fire & Rescue Service changed its role to enforcement. This was with the introduction of the Regulatory Reform (Fire Safety) Order 2005.

Fire Detection and Su28ppression Systems: Innovations like smoke detectors and automatic sprinkler systems became more common in residential and commercial buildings.

Contemporary Practices

Comprehensive Fire Codes: Modern fire codes incorporate advances in technology and understanding of fire behaviour, emphasising prevention, detection, and mitigation.

Smart Technologies: The integration of The Internet of Things (IoT) devices and AI (Artificial Intelligence) in fire safety systems enhances real-time monitoring and facilitate predictive analytics for risk assessment.

Public Awareness and Education: Modern fire safety initiatives emphasise community education, preparedness training, and the importance of fire drills in schools and workplaces.

Historical Fire Incidents and Their Impact on Fire Safety

1. The Great Fire of London (1666)

The Great Fire of London began on September 2, 1666, and lasted four days, devastating large parts of the city. The fire started in a bakery on Pudding Lane and quickly spread due to the wooden structures and narrow streets of London.

The fire destroyed 13,200 houses, numerous churches including St. Paul's Cathedral, and other significant landmarks, leaving tens of thousands homeless; amazingly, the death toll was limited to 6 people.

Consequences for Fire Safety:

Building Regulations: In response, the City of London implemented strict building codes requiring the use of fire-resistant materials like brick and stone, fundamentally changing urban architecture.

Firefighting Improvements: The incident led to the establishment of organised fire brigades and better fire response strategies.

2. The Triangle Shirtwaist Factory Fire (1911)

This tragic fire occurred on March 25, 1911, in a garment factory in New York City, claiming the lives of 146 workers, mostly young immigrant women. The fire was fuelled by flammable materials and exacerbated by locked exit doors, preventing escape. The incident highlighted appalling working conditions and the lack of safety measures in industrial workplaces.

Consequences for Fire Safety:

The fire became a catalyst for labour reforms, leading to improved workplace safety standards and regulations. New York established the Bureau of Fire Prevention, enacting fire safety codes that mandated sprinklers and improved exit access.

3. The Cocoanut Grove Fire (1942)

This nightclub fire in Boston (USA) on November 28, 1942, resulted in 492 deaths, making it one of the deadliest fires in U.S. history. The fire was sparked by a match igniting flammable decorations; overcrowding worsened the tragedy. The high death toll raised awareness about the dangers of inadequate safety measures in public spaces.

Consequences for Fire Safety:

Revised Safety Codes: Stricter fire codes for assembly buildings were enacted, including occupancy limits and automatic fire suppression systems.

Public Awareness Campaigns: Increased awareness of fire safety in crowded venues led to better emergency planning.

4. The MGM Grand Fire (1980)

The MGM Grand Hotel fire in Las Vegas on November 21, 1980, resulted in 85 deaths and over 700 injuries. The fire started due to electrical malfunctions and spread rapidly owing to the lack of sprinklers in the casino area. The incident exposed serious flaws in fire safety measures in high-rise buildings and entertainment venues.

Consequences for Fire Safety:

Sprinkler Regulations: The fire led to laws requiring sprinkler systems in all new hotels and casinos.

Emergency Planning Improvements: Enhanced training for hotel staff and emergency response protocols were established.

5. The Grenfell Tower Fire (2017)

The Grenfell Tower fire occurred on June 14, 2017 in London, claiming at least 72 lives and injuring many others. The fire started in a flat and spread rapidly due to the flammable cladding used in the building's refurbishment, that was non-compliant with safety regulations. The tragedy exposed severe deficiencies in fire safety regulations and building standards, particularly in high-rise residential buildings.

Consequences for Fire Safety:

The Grenfell tragedy prompted a comprehensive review of fire safety regulations in the UK, leading to changes in building codes, particularly regarding cladding materials and fire safety assessments. The ongoing public inquiry aims at holding those responsible for safety failures accountable and at ensuring that similar tragedies do not occur in the future.

The incident raised awareness about the importance of fire safety in social housing and the need for improved communication about risks to residents.

These historical fire incidents, including the Grenfell Tower tragedy, have significantly shaped fire safety and engineering practices. Each event highlighted critical weaknesses in existing safety measures, leading to reforms that aimed at preventing future tragedies. By learning from the past, the fire engineering profession continues to evolve, prioritising safety in design, construction, and emergency preparedness.

Chapter 3

Fire Engineering Science

Fire Engineering Science is a multidisciplinary field that applies principles of physics, chemistry, and engineering to understand, prevent, and mitigate the effects of fire. It encompasses the study of fire dynamics, combustion processes, heat transfer, fluid mechanics, thermodynamics, and material behaviour under fire conditions. This document provides a detailed examination of key concepts in fire engineering science, including relevant formulas.

While the formulas are not exhaustive, they do present the necessary formulas for fire engineers to demonstrate the basic properties of fire. There are further formulas presented throughout this document to complement this section.

FUNDAMENTALS OF FIRE SCIENCE

Fire Triangle And Fire Tetrahedron

- Fire requires three essential elements: heat, fuel, and oxygen. This is represented as the fire triangle.
- A more advanced model, the fire tetrahedron, includes a fourth component: chemical chain reactions.

Mathematically, fire can be sustained if:

Sustained Fire $\Rightarrow$ Heat + Fuel + Oxygen + Chemical Reactions

Combustion And Heat Release

Types of Combustion

Complete Combustion: Produces carbon dioxide (CO_2) and water (H_2O)

Incomplete Combustion: Produces carbon monoxide (CO), soot, and other hydrocarbons due to insufficient oxygen.

General chemical reaction for hydrocarbon combustion:

$$C_xH_y + O_2 \rightarrow CO_2 + H_2O + Heat$$

Heat Release Rate (HRR)

The rate at which energy is released during combustion is given by the following formula, it calculates the **rate at which energy is being released** during combustion, which is the driving force of fire behaviour. The formula is used in thermodynamic science and combustion science, it originates form **Lavoisier's work** (late 1700s) on energy conservation in chemical reactions.

$$\dot{Q} = \dot{m} \cdot \Delta H_c$$

Where:

$\dot{Q}$ = heat release rate (kW),

$\dot{m}$ = mass loss rate of fuel (kg/s),

ΔH_c = heat of combustion (kJ/kg).

Fire Growth And Flame Spread

Fire Growth Model

Fire growth is often modelled as a quadratic function of time in its early stages:

$$Q = at^2$$

Where:

Q = heat release rate (kW)

a = fire growth coefficient (kW/s^2)

t = time (s)

Fire growth is categorized into:

Slow: $a = 0.00293 kW/s^2$
Medium: $a = 0.0117 kW/s^2$
Fast: $a = 0.0469 kW/s^2$
Ultrafast: $a = 0.1876 kW/s^2$

Flame Height Correlation

This formula originates from experimental work by **Heskestad (1970s–1980s)**, a key figure in fire dynamics. It is often referred to as **Heskestad's Flame Height Correlation**. The relationship was derived through numerous laboratory and real-scale fire experiments.

The height of a flame depends on the HRR and can be estimated using the following,

$$L_f = 0.235 \cdot Q^{2/5}$$

Where:

L_f = flame height (m),

Q = heat release rate (kW).

Heat Transfer Mechanism

Fire spreads through three primary heat transfer mechanisms: conduction, convection, and radiation.

This is **Fourier's Law of Heat Conduction**, one of the foundational equations in thermal physics and heat transfer, particularly relevant in **fire engineering** when evaluating how heat moves through materials.

Jean-Baptiste Joseph Fourier, a French mathematician and physicist, developed this equation in the early 1800s. He published it in his groundbreaking work, *Théorie analytique de la chaleur* (*The Analytical Theory of Heat*, 1822).

Conduction

Heat transfer through a solid material is governed by Fourier's Law:

$$q = k \frac{dT}{dx}$$

Where:

q = heat flux (W/m²),

k = thermal conductivity (W/m·K),

dt/dx = temperature gradient.

Convection

This is the equation for **convective heat transfer**, often called **Newton's Law of Cooling**. It's an essential principle in fire engineering and thermal sciences.

This law was proposed by **Sir Isaac Newton** in the 1700s as an empirical relationship to describe **how objects cool** in a surrounding fluid (like air or water). It has since been refined and formalized within heat transfer theory.

$$q = h \cdot A \cdot (T_s - T_\infty)$$

Where:

h = convective heat transfer coefficient (W/m²·K),

A = surface area (m²),

T_s = surface temperature (K),

T_∞ = ambient temperature (K).

Radiation

This is the **Stefan–Boltzmann Law**, describing **radiative heat transfer**, one of the three main mechanisms of heat transfer in fire science.

The law is named after two physicists:

- **Josef Stefan** (1879) – an Austrian physicist who first proposed the empirical relationship.
- **Ludwig Boltzmann** – derived it theoretically from thermodynamic principles and blackbody radiation theory.

Radiation plays a dominant role in:

- Fire spread,
- Thermal exposure to nearby surfaces,
- Human tenability (skin burn thresholds),
- Radiant heat impact from large fires (e.g., wildfires, industrial).

When it's used:

- To estimate heat flux to adjacent buildings or materials,
- In fire growth modelling,
- For fire-resistance testing of walls, doors, and partitions,
- In analysing smoke layer radiation to structural elements.

$$q = \varepsilon \sigma T^4$$

Where:

q = radiative heat flux (W/m²),

ε = emissivity of the surface (0 to 1),

σ = Stefan-Boltzmann constant $\left(5.67 \times 10^{-8} W/m^2 K^4\right)$

T = absolute temperature (K).

Fire Dynamics And Smoke Movement
Buoyant Flow and Plume Theory

This equation estimates the **velocity of smoke rising** due to **buoyancy**, a key concept in **plume dynamics** and **smoke movement** modelling in fire engineering. This expression is based on fundamental fluid mechanics and thermodynamics, particularly from:

- The **Buoyancy Law** (Archimedes' Principle),
- Adaptations from **Froude Number scaling** in fire plume theory,
- Work by **Thomas, Zukoski, Heskestad**, and others in fire plume dynamics research, particularly during the 1960s–1980s.

It's not from a single person but rather an accumulation of research in the field of fire dynamics and mechanical ventilation modelling.

Why it matters:

Understanding buoyant smoke velocity is critical for life safety, particularly for:
- Smoke control and extraction design,
- Predicting smoke layer formation,
- Evacuation timing and egress modelling,
- Designing smoke vents, atria, and stair pressurization systems.

When it's used:
- In CFD simulations (e.g. Smartfire, FDS),
- For zonal fire models (B-Risk, CFAST),
- In the sizing of mechanical smoke control systems,
- In calculating time to untenable conditions in fire safety engineering.

$$v = \sqrt{2gH\left(T_s - T_\infty\right)T_\infty}$$

Where:

v = velocity of rising smoke (m/s),

g = gravitational acceleration (9.81 m/s²),

H = height above fire (m),

T_s = source temperature (K),

T_∞ = ambient temperature (K).

Gas Properties

Mass Fraction, Volume Fraction, Mole Fraction

The formula below represents different ways to express **gas composition**, a fundamental concept in fire chemistry and combustion science. It shows how to express the proportion of a component in a gas mixture.

These fractions are crucial when dealing with:
- Combustion gas analysis (e.g., determining levels of CO, CO_2, H_2O),
- Smoke toxicity and asphyxiant gas modelling,
- Fire modelling tools (e.g. Smartfire, FDS, CFAST),
- Oxygen consumption calorimetry (HRR calculations),
- Ventilation control in fire compartments.

When and Why It's Used
- **Mass Fraction** is used when dealing with **weight-based combustion equations** or when mass is conserved (e.g., HRR calculations).
- **Volume Fraction** is used in **gas detection systems** and **ventilation calculations**.
- **Mole Fraction** is essential in **chemical kinetics, stoichiometry**, and **equilibrium calculations**.

$$Y_i = \frac{m_i}{m_{total}} \quad X_i = \frac{v_i}{v_{total}} \quad x_i = \frac{n_i}{n_{total}}$$

Where:

Y_i = mass fraction

X_i = volume fraction

x_i = mole fraction

Average Molar Mass

$$M_{avg} = \sum x_i M_i$$

Moles and Mass

$$n = \frac{m}{M}$$

Where:

n = number of moles

m = mass (kg)

M = molar mass (kg/mol)

Ideal Gas Law

This is the Ideal Gas Law, a fundamental equation in chemistry and fire engineering that describes the behaviour of gases under various conditions.

The Ideal Gas Law is a combination of several empirical gas laws:

- **Boyle's Law** (1662) — pressure vs. volume,
- **Charles's Law** (1787) — volume vs. temperature,
- **Avogadro's Law** (1811) — volume vs. amount of gas.

It was later combined into one unified expression by 19th-century scientists, most notably **Émile Clapeyron** (1834).

Why it matters:

In fire scenarios, gases (e.g., smoke, combustion products, fresh air) are subject to extreme changes in:

- Temperature (due to heat),
- Volume (expansion and flow),
- Pressure (in sealed or ventilated compartments).

This law allows engineers to:

- Predict smoke expansion in rooms,
- Model gas concentrations and toxicity levels,
- Calculate ventilation and overpressure effects.

When it's used:

- In zone models (like B-Risk, CFAST),
- During gas flow calculations in ducts and vents,
- For estimating air entrainment in smoke plumes,
- In oxygen depletion modelling.

$$PV = nRT$$

Where:

P = pressure (Pa)

V = volume (m²)

n = moles of gas

R = gas constant

T = temperature (K)

Density

$$p = \frac{m}{v}$$

Atmospheric Air Properties

Standard composition of dry air
- 78.09% nitrogen
- 20.95% oxygen
- 0.93% argon
- 0.04% carbon dioxide

Fluid Mechanics

Bernoulli's Equation

The equation was formulated by **Daniel Bernoulli**, a Swiss physicist and mathematician, in **1738**. It reflects the **principle of conservation of energy** for flowing fluids.

Applications:
- Smoke control systems (predicting pressure and airflow through vents and doors),
- Sprinkler and hydrant system design (calculating water velocity and pressure),
- Ventilation and pressurization systems (e.g., stairwells, escape routes),
- Fire modelling of jets and plumes (CFD, empirical models).

When It's Used:
- In **Smartfire** and other CFD tools to simulate pressure and flow fields,
- When evaluating **air movement through compartments**,
- During **evacuation modelling** to assess pressure differences across doors,
- In sizing of **ducts, nozzles, and jets**.

$$P + \frac{1}{2}pv^2 + pgh = Constant$$

Where:

P = pressure (Pa)

v = velocity (m/s)

p = density (kg/m^3)

g = gravitational acceleration (9.81 m/s²)
h = height (m)

Coefficient of Contraction

The coefficient of contraction stems from classical **hydraulic theory** and was formalized through experimental work by **Torricelli** and **Venturi** in the 17th and 18th centuries. It has since become part of standard fluid dynamics and hydraulic engineering.

The coefficient tells us how much the **flow stream contracts** after exiting an orifice. This happens because the streamlines converge as the fluid exits, forming a **vena contracta**, the point where the cross-sectional area of the fluid is at its minimum.

Why it matters:

Understanding the contraction of fluid jets helps in:
- Designing fire sprinkler heads and nozzles,
- Optimizing flow discharge in fire suppression systems,
- Improving the accuracy of water stream targeting in firefighting.

When it's used:
- In sprinkler and orifice plate design,
- In vent and aperture flow modelling (e.g., in smoke control or pressure relief),
- When calculating actual **vs.** theoretical discharge rates.

$$C_c = \frac{A_{jet}}{A_{orifice}}$$

Where:

C_c = contraction coefficient

A_{jet} = jet area

$A_{orifice}$ = orifice area

DSEAR (Dangerous Substances and Explosive Atmospheres Regulations)

DSEAR ensures protection against explosive atmospheres by controlling:
1. Identification of hazardous substances
2. Risk Assessment
3. Implementing control measures
4. Emergency preparedness

This formula is used to assess flammability risk and is particularly relevant under DSEAR, the Dangerous Substances and Explosive Atmospheres Regulations.

What It Represents

- LEL = Lower Explosive Limit (or Lower Flammability Limit), expressed as a percentage (%)
- vfv = volume of fuel (vapour or gas) in air
- vtv = total volume of air–fuel mixture

This ratio expresses how close the concentration of a flammable gas or vapor is to the minimum concentration required to ignite.

Why It's Used in Fire & Explosion Risk Assessment

Why it matters:

- The LEL indicates the lowest concentration at which a gas can ignite in air.
- Below this limit, the mixture is too lean to burn.

When and where it's used:

- In DSEAR assessments for workplaces where flammable substances are handled (e.g., fuel storage, spray booths, laboratories),
- During hazard zoning for ATEX compliance (Zones 0, 1, 2),
- When specifying gas detectors and alarms,
- In determining ventilation needs for safe dilution of vapours.

$$LEL = \frac{v_f}{v_t} \times 100$$

Where:

LEL = lower explosive limit

v_f = volume of flammable gas

v_t = total volume of gas mixture

Upper and Lower Explosive Limits of Common Chemicals

Chemical/Gas	Lower Explosive Limit (LEL) [%]	Upper Explosive Limit (UEL) [%]
Methane (CH4)	5.0	15.0
Propane (C3H8)	2.1	9.5
Butane (C4H10)	1.6	8.4
Hydrogen (H2)	4.0	75.0
Carbon Monoxide (CO)	12.5	74.0
Ethanol (C2H5OH)	3.3	19.0
Benzene (C6H6)	1.2	7.8
Acetylene (C2H2)	2.5	100.0
Ammonia (NH3)	15.0	28.0
Toluene (C7H8)	1.2	7.1
Ethylene (C2H4)	2.7	36.0
Xylene (C8H10)	1.1	7.0
Acetone (C3H6O)	2.5	13.0
Gasoline	1.4	7.6
Diesel	0.6	7.5
Formaldehyde (CH2O)	7.0	73.0
Hydrogen Sulphide (H2S)	4.0	46.0
Nitromethane (CH3NO2)	7.3	62.0
Styrene (C8H8)	0.9	6.8
Hexane (C6H14)	1.1	7.5

Oxygen Depletion

Fire suppression through oxygen depletion follows:

$$O_2 \leq 15\%$$

Fires typically extinguish when oxygen levels drop below 15%, and smouldering combustion ceases around 10%.

Smoke Production Rate

This equation represents the **smoke production rate**, which is essential for modelling **smoke movement**, **tenability**, and **evacuation strategies** in fire safety engineering.

Origin and Development

This relationship has been widely studied and adopted from:

- Empirical fire testing (e.g., ISO 5660 Cone Calorimeter tests),
- Work by fire researchers like Tewarson, Babrauskas, and institutions like NIST,
- It is integrated into fire modelling tools like Smartfire, FDS and CFAST.

Why it matters:

Smoke is the primary threat to occupants in most fires. Understanding its production rate helps:

- Predict visibility loss (critical for evacuation),
- Estimate toxic gas concentrations,
- Model smoke layer descent,
- Design smoke control systems.

When it's used:

- In performance-based fire engineering design,
- In calculating Available Safe Egress Time (ASET),
- For sizing ventilation and smoke extraction systems,
- In fire simulations (Smartfire, FDS, BRANZFIRE, PyroSim).

Smoke production is estimated as:

$$\dot{m}_{smoke} = \emptyset . \dot{m}_f$$

Where:

m_{smoke} = smoke mass flow rate (kg/s),

$\emptyset$ = smoke yield factor,

m_f = fuel mass loss rate (kg/s).

$$Q = mc_p \Delta T + mL_v$$

Where:

Q = heat absorbed (J),

m = mass of water (kg),

c_p = *specific heat of water* (4.18 kJ/kg\cdotpK),

ΔT = temperature change (K),

L_v = latent heat of vaporization (2260 kJ/kg).

Structural Fire Resistance
Heat Transfer Through Walls

This is the **one-dimensional transient heat conduction equation**, also known as the **Heat Diffusion Equation**. It describes how temperature changes over time and space within a solid material like a wall during a fire.

The foundation for this equation was laid by Joseph Fourier in the early 19th century. His work on heat transfer, presented in Théorie analytique de la chaleur (1822), introduced the concept of heat conduction over time, leading to this widely used partial differential equation.

Why it matters:

It enables engineers to:

- Predict how fast heat penetrates through walls, floors, ceilings, and insulation,
- Assess thermal degradation of materials,
- Calculate the time to ignition or failure of fire barriers,
- Evaluate fire resistance performance (e.g., 30/60/120-minute ratings).

When it's used:

- In modelling heat transfer through building elements during a fire,
- For assessing compliance with BS 476, EN 1363, or ASTM E119,
- In thermal simulation software (e.g., Heat Transfer Module in Smartfire, FDS or ANSYS),
- For passive fire protection system design.

Example Application:

If you have a concrete wall exposed to fire on one side, this equation models how temperature diffuses through the wall, affecting:

Adjacent compartments,

Integrity of fire compartments,

Tenability and fire spread risk.

$$\frac{\partial T}{\partial t} = a \frac{\partial^2 T}{\partial x^2}$$

Where:

$a = k/(pc_p)$ = thermal diffusivity (m2/s),

T = temperature (K),

t = time (s),

x = position (m).

Fire Suppression Mechanism

Cooling with Water

This equation is fundamental in **fire suppression science**, particularly for understanding how **water cools a fire**. The equation derives from classical thermodynamics—not credited to one person, but built upon foundational work by scientists like:

- **James Joule** (energy conservation),
- **Rudolf Clausius** (latent heat and entropy),
- **Joseph Black** (specific and latent heat concepts, 18th century).

Two-Stage Cooling Process:

This formula breaks heat absorption by water into **two stages**:

1. **Sensible heat**:
   mcpΔT – raising water temperature from ambient to 100°C
2. **Latent heat**:
   mLv – converting water to steam, which absorbs a **huge amount of heat**

Why it matters:

- Water's ability to absorb large amounts of heat makes it an effective extinguishing agent.

- Helps quantify how much water is needed to cool materials or stop flame spread.

When it's used:
- In designing sprinkler and suppression systems,
- When calculating water demand for firefighting operations,
- To model cooling time of hot surfaces or gases,
- In evaluating fire brigade intervention effectiveness.

$$Q = mc_p \Delta T + mL_v$$

Where:

Q = heat absorbed (J),

m = mass of water (kg),

c_p = *specific heat of water* (4.18 kJ/kg\cdotpK),

ΔT = temperature change (K),

L_v = latent heat of vaporization (2260 kJ/kg).

Fire Engineering Science integrates combustion chemistry, thermodynamics, heat transfer, and fluid dynamics to analyse fire behaviour and develop fire safety strategies. Understanding formulas governing fire growth, heat release, smoke movement, and suppression mechanisms is crucial for designing effective fire protection systems.

Chapter 4

Fire Dynamics

Fire dynamics is the study of how fires start, grow, and interact with the surrounding environment. It draws on a wide range of scientific disciplines—including thermodynamics, fluid mechanics, combustion chemistry, and materials science—to explain the complex phenomena that occur during fire events. This field investigates not only the physical processes involved in ignition, flame spread, and heat transfer but also the impact of building geometry, ventilation, and material properties on fire behaviour.

Understanding fire dynamics is essential for fire safety professionals, engineers, architects, and emergency responders. By analysing the variables that affect fire growth, such as fuel load, ventilation conditions, and heat release rate, professionals can anticipate fire development and make informed decisions about fire prevention, detection, and suppression strategies. Furthermore, fire dynamics informs the design of fire protection systems, such as sprinklers, smoke control, and fire-resistant construction, as well as egress planning and occupant safety protocols.

Modern computational tools, such as Smartfire and other modelling software, have further advanced our understanding of fire behaviour by allowing the simulation of complex scenarios in a virtual environment. These insights help assess the performance of buildings under fire conditions and guide the development of performance-based fire safety designs.

In essence, fire dynamics forms the scientific backbone of fire engineering, enabling the prediction and control of fires to safeguard life, property, and the environment.

Stages of Fire Growth

1. Ignition (Incipient Stage)

The ignition or incipient stage represents the birth of a fire event and is defined by the initiation of combustion in a fuel source. This occurs when a material reaches its ignition temperature and a sufficient heat source provides the necessary energy to break molecular bonds, enabling a sustained chemical reaction with oxygen.

- Ignition sources are varied and can include open flames (such as matches or candles), electrical arcing or faults, hot surfaces, chemical reactions (such as spontaneous combustion), or mechanical friction.
- Combustion at this point is localized and typically incomplete, with high oxygen availability and low levels of heat and smoke production.
- The fire is generally small and stable, allowing for potential self-extinguishment or early intervention through manual suppression (e.g., extinguishers or water application).
- Physical indicators may include slight discoloration or charring of materials, a faint smell of smoke, or small visible flames depending on the fuel and environment.
- In enclosed spaces, temperature rise is negligible, and visibility is largely unaffected.

The incipient stage is particularly critical in fire safety because it offers the greatest opportunity for intervention before the fire escalates. Effective detection systems, such as smoke detectors or heat sensors, can identify this stage early, triggering alarms and enabling occupants to evacuate or suppress the fire.

From a chemical perspective, this stage marks the beginning of pyrolysis in solid fuels, where thermal decomposition releases flammable vapours. These vapours mix with oxygen in the surrounding air and, given an adequate ignition source, begin to sustain a flame.

Understanding the factors that contribute to ignition, such as ignition energy, surface temperature, fuel moisture content, and ventilation, is fundamental to fire prevention strategies. Proper maintenance of electrical systems, control of ignition sources, and use of flame-retardant materials are all measures that can prevent a fire from entering the subsequent growth phase.

2. Growth Stage

The growth stage represents a critical period in fire development during which the fire evolves from a localized flame to a rapidly expanding and more destructive phenomenon. This phase is characterized by a substantial increase in the rate of combustion, heat release, and the spread of flames.

- As the fire consumes additional fuel and draws in oxygen, it begins to accelerate. The heat produced increases the ambient temperature within the compartment, which can cause nearby combustible materials to undergo pyrolysis and emit flammable vapours.

- The heat released during this stage is sufficient to preheat adjacent fuel surfaces, creating conditions that favour flame spread both vertically (towards ceilings and upper levels) and horizontally (across walls, floors, and furnishings).
- Smoke production becomes more pronounced as materials begin to degrade, and combustion remains partially incomplete. This smoke may include a mixture of hot gases, particulates, and toxic substances such as carbon monoxide (CO), hydrogen cyanide (HCN), and acrolein, which pose severe health risks.
- The visual environment deteriorates rapidly, visibility can be reduced to near zero within seconds, making escape difficult and disorienting for occupants.
- Key environmental factors, such as the geometry of the space, presence of synthetic furnishings, availability of ventilation openings (doors, windows, HVAC), and thermal properties of structural materials, significantly influence the speed and direction of fire growth.
- Uncontrolled ventilation, such as an open window or door, can feed oxygen into the compartment and accelerate combustion, a phenomenon known as ventilation-controlled burning.
- A pivotal event that may occur during this stage is **flashover**. As heat and combustible gases accumulate, surfaces throughout the enclosure may reach ignition temperature nearly simultaneously, transitioning the fire into its most intense and dangerous phase.
- The growth stage is the last phase during which some form of evacuation or suppression remains feasible. Failure to act quickly can result in rapid escalation beyond safe intervention.

During this stage, fire dynamics principles are particularly relevant for predicting fire growth and for planning effective intervention strategies. Tools such as time-temperature curves, plume models, and computational simulations help engineers and emergency responders estimate how quickly a fire may develop, allowing them to design systems that improve occupant survivability and delay flashover.

3. Flashover (Critical Transition Phase)

Flashover is one of the most critical and dangerous transitions in the lifecycle of a compartment fire. It marks the moment when localized burning evolves into full-room involvement, where nearly all exposed combustible materials ignite almost simultaneously. This sudden escalation in fire intensity transforms the environment into an unsurvivable space within seconds.

- Flashover generally occurs when upper layer gas temperatures within a room exceed approximately 500–600°C. At this temperature, radiative heat transfer from the hot smoke layer becomes intense enough to ignite surface materials throughout the compartment.
- This process is heavily influenced by heat feedback mechanisms: as burning materials emit radiant energy, surrounding surfaces are preheated, releasing additional pyrolysis gases. These gases, now in sufficient quantity and uniformly distributed, ignite in a cascading effect.
- Warning signs of impending flashover include a rapid increase in room temperature, rollover or ghosting flames (small tongues of flame that dance along the smoke layer), dense smoke lowering from the ceiling, and the spontaneous ignition of objects far from the flame front.
- During flashover, visibility drops to zero, air temperatures reach lethal levels, and survival for any occupants inside becomes nearly impossible without protective gear and immediate egress.
- Firefighter safety protocols emphasize the identification of pre-flashover conditions to avoid becoming trapped in a compartment that transitions without warning.

From a fire science perspective, flashover is considered the boundary between the growth and fully developed stages. It also represents the limit of effectiveness for many detection and suppression systems. Fire modelling tools, such as Smartfire, use temperature thresholds, heat release rate (HRR), and surface conditions to predict flashover potential in performance-based design scenarios.

Understanding flashover dynamics is critical in:
- Planning effective compartmentation and passive fire protection.
- Establishing safe egress timelines.
- Designing early intervention systems that can delay or prevent flashover, such as quick-response sprinklers and smoke ventilation systems.

4. Fully Developed Stage

The fully developed stage represents the peak of a fire's intensity and destructive capability. It is characterized by the complete involvement of all available combustible materials within a compartment, resulting in the highest levels of heat release, flame spread, and structural impact.

- During this phase, the fire is no longer limited by the availability of fuel but instead often becomes ventilation limited. The combustion rate is primarily dictated by the amount of oxygen that can be supplied to sustain the reaction.
- Temperatures typically exceed 1000°C, and the thermal environment becomes completely untenable for human survival. Even firefighting equipment and protective gear may reach their performance limits.
- The flames may engulf the entire space, emerging from windows or openings as they seek fresh air, intensifying the burn.
- Structural components such as steel beams or concrete may begin to weaken or fail due to sustained thermal loading. Wood-based materials may char and eventually collapse.
- Radiant heat becomes the dominant mode of heat transfer, potentially igniting adjacent compartments or external exposures through window openings or façade systems.
- Firefighters must exercise extreme caution during this phase, as sudden ventilation (such as breaking windows or opening doors) can trigger a backdraft, an explosive ignition of superheated gases when oxygen is suddenly introduced.
- Strategic suppression operations, such as transitional attack methods (cooling the environment before entering) and use of compartmentalization, are critical to control the fire without exacerbating risks.

In fire engineering, this stage is crucial for evaluating fire-resistance ratings of materials and assemblies. Time-temperature curves (such as those used in BS 476 or EN 1363) simulate this stage to determine how long a structure can withstand intense fire exposure before failure.

Understanding the dynamics of the fully developed fire enables professionals to:
- Establish appropriate fire resistance levels for load-bearing structures.
- Design effective compartmentation to prevent fire spread.
- Specify suitable suppression systems, like sprinkler densities and nozzle types.
- Plan emergency response strategies that prioritize firefighter safety and structural stability.

5. Decay Stage

The decay stage is the final phase in the fire development cycle, marking a decline in fire intensity due to the depletion of one or more essential elements, typically fuel or oxygen. While often perceived as a less hazardous period, this stage can still pose significant risks to both occupants and emergency responders.

- As available fuel is consumed and oxygen becomes scarce, the rate of combustion decreases, leading to a gradual reduction in temperature and visible flames.
- However, smouldering combustion may continue within insulated or concealed areas, such as upholstered furniture, wall cavities, or ceiling voids. These hidden hotspots can persist for hours and are difficult to detect without specialized equipment.
- Despite the reduction in open flames, the environment may remain extremely hazardous. High levels of carbon monoxide (CO), a colourless, odourless, and toxic gas, can accumulate, presenting serious risks of asphyxiation to anyone exposed without respiratory protection.
- In some cases, re-ignition or rekindling can occur if oxygen is reintroduced to a smouldering fire. This is especially dangerous in ventilation-limited fires where sudden openings, like a broken window or opened door, may lead to a rapid resurgence in fire activity.
- Structural stability remains a concern during this phase. Prolonged exposure to high temperatures may have weakened load-bearing elements, making collapse more likely even as the fire appears to subside.
- Firefighters must remain vigilant during overhaul operations, using thermal imaging cameras and other tools to locate hidden embers and ensure complete extinguishment.

From a fire dynamics perspective, the decay stage highlights the importance of continuous monitoring and cautious re-entry into post-fire environments. It also emphasizes the need for effective smoke management, ventilation, and post-fire inspection to mitigate lingering hazards.

Fire Development Curve

The fire development curve is a fundamental concept in fire dynamics, illustrating the typical progression of a fire through its various stages: ignition, growth, flashover, fully developed, and decay. This curve helps visualize how quickly a fire can intensify and underscores the importance of early detection and intervention.

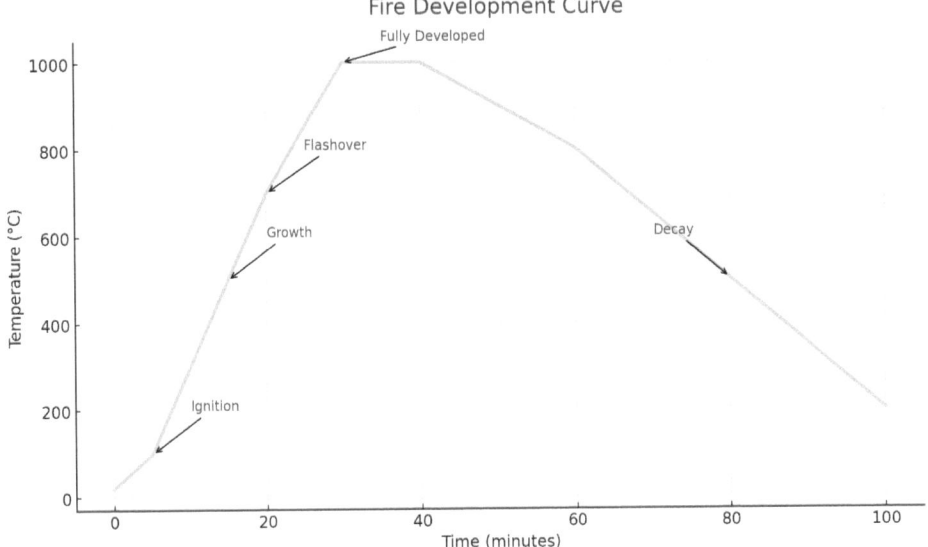

- **X-axis (Time)**: Represents the duration since ignition, usually in minutes.
- **Y-axis (Temperature)**: Represents the temperature within the fire compartment in degrees Celsius (°C).

Each segment of the curve corresponds to a distinct stage:

- **Ignition**: A slow rise in temperature as the fire begins.
- **Growth**: Rapid increase in temperature as the fire spreads and consumes more fuel.
- **Flashover**: A sudden spike when the environment becomes fully involved—nearly all materials ignite simultaneously.
- **Fully Developed**: The peak of the fire, characterized by maximum heat release and structural damage.
- **Decay**: A decline in temperature as fuel and oxygen are exhausted.

This curve is critical for:

- Estimating time available for safe evacuation.
- Designing suppression systems and fire-resistant materials.
- Training emergency responders on the stages of fire progression.

Stage	Temperature Range	Key Characteristics	Risks & Responses
Ignition	~20–100°C	Small flame, low HRR, minimal smoke	Early detection and manual extinguishment
Growth	100–600°C	HRR increases, smoke develops, flashover risk	Initiate evacuation, activate suppression
Flashover	500–700+°C	Total involvement, rapid fire spread	Immediate evacuation, system saturation
Fully Developed	800–1100°C	Peak HRR, structural threat, backdraft risk	Firefighting from protected positions
Decay	Cooling down	Fuel/oxygen depletion, CO risk, instability	Overhaul and hotspot monitoring

Heat Transfer Methods

Fire spreads primarily through three methods: conduction, convection, and radiation.

Conduction

Conduction is the transfer of thermal energy through a solid medium as a result of molecular interaction. When one part of a material is heated, its molecules begin to vibrate more rapidly. These vibrations are transferred to adjacent molecules, propagating heat through the material without any bulk movement of the substance itself. This form of heat transfer plays a significant role in the spread of fire, especially within structural elements and building materials.

- Conduction is most efficient in materials with high thermal conductivity, such as metals. In the context of fire, metal beams, pipes, and conduits can rapidly conduct heat from one area to another, potentially igniting combustible materials at some distance from the original flame source.
- For example, in steel-framed buildings, heat can travel along beams and cause distant areas of a structure to weaken or fail before the flames even reach them. Likewise, hot gases in ducts or pipes can raise the temperature of enclosures or walls far removed from the fire origin.
- In composite structures or layered assemblies (e.g., walls and floors), conduction through fasteners, bolts, and internal reinforcements can serve as unintended ignition pathways, especially if insulation is inadequate.

- The rate of heat transfer by conduction depends on the material's thermal conductivity (k), the temperature gradient across the material (dT/dx), and the cross-sectional area through which heat flows. This relationship is governed by Fourier's Law:

q = -k (dT/dx)

where:
  - q is the heat flux (W/m²),
  - k is the thermal conductivity of the material (W/m·K),
  - dT/dx is the temperature gradient (K/m).

- The negative sign in the equation indicates that heat flows from regions of higher temperature to regions of lower temperature.
- In fire safety engineering, conduction must be considered in the design of fire barriers, insulation systems, and passive fire protection. Materials with low thermal conductivity (e.g., gypsum board, mineral wool) are preferred to slow heat propagation and compartmentalize fires effectively.

Recognizing the role of conduction in fire spread helps engineers anticipate potential fire growth paths, especially in concealed or interconnected spaces. It also informs decisions on material selection and the placement of thermal breaks to prevent unintended ignition beyond fire-rated zones.

Convection

Convection is the transfer of heat through the movement of fluids, either gases or liquids, driven by temperature differences. In the context of fire dynamics, convection is especially significant because it dictates how heat and combustion by-products such as smoke and hot gases move through and affect a building environment.

- In fires, hot air and gases generated at the source rise due to their lower density, creating upward currents that can spread heat vertically through stairwells, shafts, and open spaces. This upward movement can ignite combustible materials located above the fire origin, such as ceiling tiles, insulation, or materials on upper floors.
- As the heated gases move, they displace cooler air, creating circulation patterns that can transport smoke and toxic fumes to other parts of a structure. This process, called smoke stratification, can significantly reduce visibility and tenability for occupants.
- Convection can also lead to the development of ceiling jets, high-temperature, fast-moving layers of gas that spread laterally just beneath the ceiling. These

jets are responsible for activating sprinklers and heat detectors and for causing thermal degradation in overhead materials.
- In confined or enclosed environments, the accumulation of hot gases at the ceiling level can lead to pressure build-up and thermal layering, which pose risks for flashover or backdraft if ventilation changes occur.

The rate of heat transfer through convection is influenced by the fluid properties, surface area, and temperature difference between the surface and surrounding air. This process is mathematically described by Newton's Law of Cooling:

$$q = h \times A \times (T_s - T_\infty)$$

where:

q is the convective heat transfer (W),

h is the convective heat transfer coefficient (W/m²·K),

A is the surface area exposed to the fluid (m²),

T_s is the surface temperature (K),

T_∞ is the ambient temperature (K).

The value of h varies depending on whether the fluid flow is natural (buoyancy-driven) or forced (mechanically induced), and whether the flow is laminar or turbulent.

In fire engineering, understanding convection is essential for designing smoke control systems, modelling fire growth, and predicting heat exposure to structural components.

Ventilation design, such as mechanical extraction or pressurization systems, often relies on an accurate assessment of convective heat and smoke movement to protect escape routes and enable safe evacuation.

Radiation

Radiation is the transfer of heat energy through electromagnetic waves, typically in the infrared spectrum. Unlike conduction and convection, which require a medium (solid or fluid), radiation can occur across a vacuum and is not dependent on direct contact or fluid movement.

- Radiant heat transfer is particularly significant in fire scenarios because it can ignite materials at a distance from the flame source, contributing to the rapid horizontal and vertical spread of fire.
- For example, radiant energy from a fully developed compartment fire can pass through windows or doorways and ignite materials in adjacent rooms or even across streets in wildland-urban interface fires.

- Radiation intensity increases sharply with temperature. As flames and hot surfaces reach extreme temperatures, the energy radiated can be sufficient to ignite nearby combustibles without flame contact.

Radiative heat transfer is governed by the **Stefan-Boltzmann Law**, which expresses the rate of heat emission from a surface as:

$$q = \varepsilon \times \sigma \times T^4$$

where:

q is the radiative heat flux (W/m²),

ε is the emissivity of the surface (0–1),

σ is the Stefan-Boltzmann constant (5.67×10^{-8} W/m²·K⁴),

T is the absolute temperature of the surface (K).

Emissivity varies depending on the surface material and condition. For example, polished metals have low emissivity (reflective), whereas charred wood or black surfaces have high emissivity (absorptive).

Radiative heat transfer becomes dominant in post-flashover fires and in open fires where large flame surfaces are exposed.

In fire engineering, radiation is a critical factor in:
- Predicting the spread of fire between compartments or buildings,
- Determining safe separation distances between fuel sources,
- Assessing risk of ignition for façade systems and adjacent materials,
- Designing barriers and fire-resisting glazing that can reflect or absorb radiated heat.

Understanding radiation helps fire engineers evaluate thermal exposure risks, design protective shielding, and incorporate fire-retardant treatments for materials likely to be exposed to intense heat.

FIRE DYNAMICS IN DIFFERENT ENVIRONMENTS

Compartment vs. Open-Air Fires

Fire behaviour varies significantly between enclosed compartments and open-air settings:
- In compartment fires, heat and smoke are contained, leading to rapid temperature rise and increased likelihood of flashover due to limited ventilation and heat feedback.

- In open-air fires, heat dissipates more readily, making flashover less likely, but flame spread may be more dependent on wind, topography, and fuel distribution.
- Fire dynamics in compartments are influenced by factors like room size, thermal insulation, and ventilation openings, while open-air fires are more influenced by environmental conditions.

Understanding these differences is crucial for designing appropriate fire protection systems and emergency response strategies.

Influence of Suppression on Fire Dynamics

Fire suppression measures, particularly water-based systems, alter fire dynamics by:

- Rapidly reducing temperatures and delaying or preventing flashover.
- Disrupting pyrolysis through surface cooling.
- Limiting the availability of combustible gases and reducing flame spread.
- Changing the flow paths of heat and smoke.

Inert gas systems and chemical agents act differently by smothering the fire or interrupting chemical reactions rather than cooling.

Effective integration of suppression strategies with an understanding of fire dynamics ensures faster control, less damage, and safer evacuation.

Interdisciplinary Relevance

Fire dynamics is intrinsically linked to multiple areas of fire safety engineering:

- **Law and Regulation**: Guides design standards and safety codes.
- **Material Science**: Supports selection of flame-retardant or fire-resisting materials.
- **Occupant Behaviour**: Shapes egress planning and human response modelling.
- **Mechanical Systems**: Informs HVAC, smoke control, and suppression design.

This interdisciplinary nature makes fire dynamics a foundational chapter in any comprehensive fire engineering approach.

Lessons Learned

- Fire growth can be rapid and unpredictable.

- Understanding heat transfer mechanisms helps design safer buildings.
- Flashover is a critical point that dictates evacuation timing.
- Material selection, smoke control, and ventilation strategies are life-saving interventions.

Case Studies in Fire Dynamics

Grenfell Tower Fire (2017, UK)

The Grenfell Tower fire stands as one of the most significant fire disasters in modern UK history, highlighting profound failures in building design, material selection, regulatory enforcement, and emergency preparedness. The fire began in the early hours of June 14, 2017, in a fourth-floor flat of the 24-storey Grenfell Tower residential building in North Kensington, London. An electrical fault in a fridge-freezer initiated the blaze, which rapidly developed and breached the apartment window, igniting the external cladding system.

- The external cladding consisted of aluminium composite material (ACM) panels with a polyethylene core, highly combustible and known to be a fire risk. Once ignited, the cladding facilitated vertical fire spread up all four sides of the building.
- The configuration of the rain-screen cladding, insulation (polyisocyanurate), and ventilation cavity created a 'chimney effect' that rapidly transported flames and hot gases upwards and horizontally across the façade.
- The dominant heat transfer mechanisms involved were convection (within the cavity space and up the façade) and radiation (from flames against the façade and into neighbouring flats through windows).
- Internal flashovers occurred repeatedly in individual flats as heat and flames penetrated through window openings or cladding voids. This phenomenon significantly complicated evacuation efforts and hindered firefighter operations.
- The fire spread from the fourth floor to the roof in just over 15 minutes and engulfed the entire building within hours, demonstrating an unprecedented failure of passive fire protection measures.
- Compartmentation, normally relied upon to confine fire to a single flat, was breached by the façade system, allowing fire and smoke to spread across compartments, vertically and laterally.
- Smoke and heat rapidly filled communal areas and escape routes, overwhelming the building's single stairwell, which served as the sole means of egress. The

absence of functional smoke control systems and the lack of sprinklers further exacerbated the situation.
- Tragically, 72 people lost their lives, many of them due to smoke inhalation and entrapment. Hundreds more were injured or displaced.

The Grenfell disaster triggered a national inquiry and widespread public scrutiny. It led to fundamental changes in UK fire safety legislation, including:
- A ban on combustible cladding materials in high-rise buildings,
- Strengthened Building Regulations (Approved Document B),
- Mandatory inclusion of sprinklers in new residential buildings over 11 metres,
- Reform of the Fire Safety Order to clarify responsibilities for building owners and managers,
- Establishment of a new Building Safety Regulator to oversee compliance.

From a fire dynamics perspective, Grenfell Tower exposed the catastrophic consequences of failing to consider the behaviour of new materials and construction systems under fire conditions. It underscored the importance of:
- Understanding façade fire performance,
- Designing for real-world fire spread scenarios,
- Ensuring redundancy in fire safety systems,
- Bridging the gap between fire science and regulatory practice.

Bradford City Stadium Fire (1985, UK)

The Bradford City Stadium fire occurred on May 11, 1985, during a football match at Valley Parade Stadium in Bradford, West Yorkshire. This disaster resulted in the deaths of 56 spectators and injuries to more than 250 others, highlighting the extreme danger posed by combustible materials, delayed evacuation, and the lack of fire safety infrastructure in public venues.
- The fire broke out underneath the wooden stand when a discarded cigarette or match ignited accumulated litter, including paper and plastic, which had built up over the years in a cavity beneath the seating area.
- Initially unnoticed, the fire rapidly spread through the dry, combustible debris, and within minutes engulfed the entire wooden structure of the stand.
- The primary mechanisms of heat transfer were convection, which allowed hot gases to rise through the tiered seating and exit routes, and radiation, which ignited wooden beams, plastic seating, and clothing.

- Within less than four minutes, the fire had escalated to flashover conditions within the enclosed stand, producing thick, toxic smoke that severely hindered evacuation.
- Escape routes were blocked by metal fencing and locked turnstiles, which were standard at the time to control crowd movement. Many exits were also narrow and inadequately signed, contributing to the high casualty rate.
- The stadium did not have fire extinguishers or a fire alarm system, and firefighters had limited access due to the configuration of the structure and dense smoke.

The Bradford fire prompted a major review of safety standards in sports venues and led to the Popplewell Inquiry, which made several key recommendations:

- Prohibition of new construction using timber or combustible materials in spectator stands,
- Requirement for the removal of combustible litter and routine inspection of void spaces,
- Mandated installation of fire detection and suppression systems,
- Improvements to exit design and crowd control features to facilitate safe evacuation.

From a fire dynamics standpoint, this incident demonstrated:

- How small ignition sources can trigger devastating fires when fuel and oxygen are readily available,
- The importance of good housekeeping and the elimination of combustible debris,
- The speed with which fires can progress through enclosed spaces filled with flammable materials,
- The need for fire safety systems and clear egress paths in venues with large public occupancies.

The legacy of the Bradford Stadium fire remains embedded in UK fire safety regulation, influencing future revisions to the Guide to Safety at Sports Grounds and the broader approach to managing fire risk in public assembly areas.

Station Nightclub Fire (2003, USA)

The Station Nightclub fire occurred on February 20, 2003, in West Warwick, Rhode Island, and remains one of the deadliest nightclub fires in U.S. history. The fire began when pyrotechnic devices used during a band performance ignited flammable acoustic foam lining the stage area. The highly combustible polyurethane foam quickly caught fire, creating a rapidly spreading flame front that engulfed the nightclub within minutes.

- The acoustic foam, intended for soundproofing, lacked flame-retardant properties. Once ignited, it generated thick, toxic smoke and an intense radiant heat flux that contributed to a flashover within approximately 90 seconds.
- The primary mechanisms of heat transfer were radiation, which preheated nearby materials and caused surfaces to ignite, and convection, which carried flames and hot gases through the club's open layout.
- Within minutes, the fire overwhelmed the room, creating a high-temperature, low-visibility environment that quickly incapacitated occupants. Most of the 100 fatalities occurred due to smoke inhalation and thermal injuries, especially near the main exit where congestion created a fatal bottleneck.
- Video footage taken from inside the club during the initial moments of the fire provided rare and powerful insight into how quickly a small fire can grow to life-threatening proportions. It also documented crowd behaviour and the immediate impact of toxic smoke.
- Compounding the tragedy was the lack of adequate fire suppression and alarm systems. The building did not have an automatic sprinkler system, nor were exit paths clearly marked or appropriately distributed for the occupancy level.

The fire prompted major changes to fire and building codes across the United States, including:

- Requirement for automatic sprinklers in nightclubs and similar assembly occupancies with more than 100 occupants,
- Tighter regulation on the use of combustible acoustic materials,
- Enhanced enforcement of occupancy limits,
- Improvements in crowd management and exit signage standards.

From a fire dynamics standpoint, the Station Nightclub fire underscored:

- The importance of controlling ignition sources in proximity to flammable materials,
- The deadly speed at which flashover can occur in confined public spaces,
- The necessity of early suppression systems to halt fire spread,
- The role of building layout and egress design in occupant survivability.

MGM Grand Fire (1980, USA)

The MGM Grand Fire occurred on November 21, 1980, at the MGM Grand Hotel and Casino (now Bally's Las Vegas) in Nevada, USA. It remains one of the deadliest high-rise hotel fires in U.S. history, resulting in 85 fatalities and over 700 injuries. The fire revealed numerous critical weaknesses in fire safety systems and became a defining moment in fire protection reform within the hospitality industry.

- The fire originated in a restaurant located on the casino level, caused by an electrical ground fault in a wall-mounted refrigerated pastry display case. The materials within the wall space, wood framing, insulation, and plastic piping, were highly combustible, and there were no automatic suppression systems in place.
- The fire grew rapidly and was able to spread horizontally due to the lack of fire barriers and unprotected penetrations in walls and ceilings.
- Smoke and hot gases were drawn vertically through the building via the elevator shafts, stairwells, and ventilation ducts. This movement of smoke, aided by convection, quickly reached the upper hotel floors despite the fire being confined to the lower levels.
- Conduction also played a role, with heat transferring through building materials and infrastructure, spreading fire to concealed spaces.
- Most fatalities were not due to burns, but rather to smoke inhalation. Victims on upper floors succumbed to toxic gases and oxygen depletion as smoke spread unchecked through HVAC systems and poorly compartmentalized areas.
- The fire was largely extinguished within a few hours, but the toll was severe due to inadequate detection, suppression, and evacuation provisions.

This tragedy prompted sweeping fire safety reforms in hotels and similar occupancies, including:

- Mandatory installation of fire sprinklers in all guest rooms and public spaces,
- Enhanced smoke detection and alarm systems throughout high-rise buildings,
- Compulsory smoke dampers in HVAC systems to limit smoke spread,
- Upgraded fire-resistance standards for walls, ceilings, and furnishings,
- Improved emergency lighting and stairwell pressurization systems.

From a fire dynamics perspective, the MGM Grand Fire highlighted several vital lessons:

- The rapid vertical spread of smoke and gases through building systems can be more deadly than the flames themselves.

- The absence of automatic sprinklers and smoke control measures can turn a localized ignition source into a widespread disaster.
- Effective compartmentation, early detection, and suppression are critical in preventing high-rise fire casualties.

This incident served as a major case study in fire engineering curricula and influenced codes such as NFPA 101 (Life Safety Code) and local building regulations nationwide.

Lessons Learned

- Fire growth can be rapid and unpredictable, often escalating from a small ignition source to an uncontrollable blaze in a matter of minutes. This highlights the critical need for early detection, rapid response, and proper occupant training.
- Understanding heat transfer mechanisms, conduction, convection, and radiation, enables fire engineers to predict how a fire will spread within and beyond a compartment. This knowledge informs the layout of passive fire protection and placement of structural elements to resist fire progression.
- Flashover represents a sudden, lethal transition in fire behaviour that drastically reduces tenability within a compartment. Recognizing the signs and triggers of flashover is vital for evacuation planning, suppression system activation, and firefighter safety.
- Material selection plays a crucial role in determining how quickly a fire grows and how much smoke and toxic gas is produced. Choosing non-combustible or fire-retardant materials, combined with well-designed smoke control and ventilation systems, can significantly delay fire spread and preserve escape routes.
- The integration of fire dynamics principles into building design, code enforcement, and emergency procedures strengthens resilience against fire incidents. Multidisciplinary collaboration among architects, engineers, regulators, and first responders ensures a holistic approach to life safety.
- Understanding heat transfer mechanisms helps design safer buildings.
- Flashover is a critical point that dictates evacuation timing.
- Material selection, smoke control, and ventilation strategies are life-saving interventions.

Chapter 5

Smoke Movement and Management

Smoke movement and toxicity are critical considerations in fire engineering, as they significantly impact life safety, building design, and emergency response strategies during a fire incident. Smoke management is a critical aspect of fire safety in buildings, designed to control and mitigate the movement of smoke during a fire incident. Since smoke is often the leading cause of injury and fatalities in fires, its proper management plays a vital role in protecting occupants, reducing property damage, and assisting emergency responders.

In this chapter, we will discuss, at length, the following aspects of smoke management:

- Components of Smoke Management
- Axisymmetric Plume Management
- Balcony Spill Plume Management
- Window Plume Management
- Communicating Space Smoke Plume Management
- Management of Smoke Flow from Smoke Layer
- Management of Airflow to Control Smoke Flow from Plume

Components of Smoke Management

Smoke Movement

Understanding how smoke behaves within enclosed spaces is crucial. Factors like buoyancy, stack effect, and airflow influence smoke movement.

Smoke Control Systems:

These systems aim to prevent smoke spread, maintain tenable conditions, and protect escape routes.

Active Methods: Examples include smoke purge systems, pressurization control, and smoke reservoirs.

Passive Methods: These involve fire-resistant barriers, smoke barriers, and compartmentalization.

Design Considerations

Plume Models: Estimating smoke production rates using plume models helps design effective smoke control systems.

Vertical Smoke Transportation: Understanding how smoke rises and spreads vertically is essential for designing smoke management systems.

Ceiling Jet: Addressing the movement of smoke near the ceiling level.

Smoke Spreading Under Flat Ceilings: Strategies to prevent smoke accumulation in large open spaces.

Risk Assessment and Mitigation

- Evaluate the risks associated with smoke movement in different building types (e.g., high-rise buildings, atria, malls).
- Calculate the implications of smoke stratification and reservoir size.
- Implement appropriate smoke control measures based on risk analysis.

Collaboration and Standards

- Fire engineers, architects, and building designers collaborate to integrate smoke management into building designs.
- Adherence to relevant codes and standards (such as ASHRAE guidelines, or the UK Smoke Control Association) ensures effective smoke control.

A well-designed smoke management system not only protects lives but also supports efficient firefighting efforts.

In the UK, the requirement for an Automatic Opening Vent (AOV) at the top of a stairwell in a residential building is not solely determined by the building's height. Instead, it depends on factors such as the building's design, the length of escape routes, and the overall fire safety strategy. According to Approved Document B of the Building Regulations, a smoke ventilation system, which may include AOVs, is typically required if the travel distance between an apartment door and the nearest escape exit exceeds 7.5 metres.

Additionally, for buildings with a single stairway, especially those exceeding certain heights, providing smoke ventilation to common escape routes is essential to ensure safe egress during a fire. This often involves installing AOVs at the top of stairwells to facilitate the efficient removal of smoke.

The choice between natural smoke ventilation systems, such as Automatic Opening Vents (AOVs), and mechanical smoke ventilation in the UK is influenced by the building's height and design. Natural smoke ventilation is typically suitable for buildings up to 30 meters tall. For buildings exceeding this height, mechanical smoke ventilation systems are generally required to effectively manage smoke control.

This approach aligns with the recommendations in Approved Document B of the Building Regulations, which provides guidance on fire safety matters, including smoke control measures. While natural ventilation methods are often adequate for shorter buildings, taller structures necessitate mechanical systems to ensure efficient smoke extraction and maintain safe egress routes during a fire. When deciding on which system is suitable, it is necessary to understand the movement of smoke in a fire situation.

Lithium Battery fires

Lithium battery fires are increasing due to the popularity of solar panels. The smoke produced by a fire involving lithium batteries is particularly dangerous and requires consideration when deciding on the type of smoke extraction facilities in a building.

The smoke produced contains the following.

- **Hydrogen Fluoride (HF)** – One of the most dangerous by-products; even in small quantities, it is highly corrosive and toxic to humans.
- **Carbon Monoxide (CO)** – A deadly asphyxiant that interferes with oxygen transport in the body.
- **Carbon Dioxide (CO_2)** – Contributes to oxygen depletion in enclosed spaces.
- **Hydrogen Cyanide (HCN)** – A highly toxic gas that affects the nervous system and respiration.
- **Sulphur Dioxide (SO_2) and Other Acidic Gases** – Formed due to breakdown of electrolyte solvents and additives.
- **Volatile Organic Compounds (VOCs)** – Toxic chemicals released from electrolyte decomposition.
- In addition, the smoke from lithium-ion battery fires has the following properties compared to fires from traditional fuels (e.g., wood, plastics),:
  - Produce a higher concentration of toxic gases per unit mass.
  - Emit thicker, darker smoke due to the organic solvents in electrolytes.

While the smoke does tend to be buoyant due to the hot gases, Hydrogen fluoride (HF), carbon dioxide (CO_2), and heavy metal aerosols are denser than air and can settle at lower levels. For this reason, mechanical smoke extraction should be considered on all buildings containing lithium battery banks irrelevant of the building's height.

Note: When applying water to lithium battery fires extreme care should be taken as water will turn the Hydrogen Fluoride HF into Hydrofluoric Acid $HF(aq)$ which is an extremely dangerous and highly corrosive acid. In addition to this when considering water mist as an alternative to sprinklers, extreme care should be taken where a bank of lithium batteries are present as with water mist, because of the high surface area to volume ratio, would probably be capable of forming a fine, potentially almost invisible cloud of $HF(aq)$ which would be a hazard to FRS and members of the public.

Axisymmetric Plume Management

Atrium Smoke Exhaust System with an Axisymmetric Plume

An atrium smoke exhaust system is a critical component in fire safety engineering for large open spaces like atriums. The system is designed to manage smoke in the event of a fire, ensuring safe evacuation and minimizing smoke damage. When considering an axisymmetric plume, it is essential to understand its formation and behaviour, as this affects the design and operation of the exhaust system.

Axisymmetric Plume Characteristics

An axisymmetric plume occurs when a fire source produces buoyant smoke that rises symmetrically around its vertical axis. This type of plume is typical for point sources of fire (e.g., a fire in the centre of a room or open space). Key characteristics include:

- *Symmetry*: The plume spreads radially outward as it rises.
- *Temperature and velocity gradients*: These decrease with height due to entrainment of surrounding air.
- *Buoyancy-driven flow*: The hot gases rise because of their lower density compared to the surrounding air.

Smoke Layer Formation in Atriums

If a fire occurs and continues to develop on a storey directly open to an atrium, hot smoke rises to the ceiling level of that storey and spreads outwards from the fire to form a layer beneath the ceiling. If that storey opens directly onto an atrium, smoke flows from the ceiling layer into the atrium void, where it tends to rise upwards owing to its buoyancy. As the smoke rises through the atrium it entrains large quantities of cool air from its surroundings, reducing the temperature of the plume and increasing its mass and volume. As the smoke plume rises, it cools, and its buoyancy reduces to such an extent that at some height its temperature can fall to that of the surrounding air, and it ceases to rise by its own buoyancy.

In such circumstances, a stable layer of smoky gases can form some distance below the atrium roof. Having risen to an upper limit, the smoke then tends to build downwards, producing a layer of increasing depth which spreads horizontally into any open storeys within the depth of the layer. This can happen even with AOV's located at the top of the atrium.

- *Plume development*: The smoke rises vertically until it encounters the ceiling.
- *Ceiling jet*: The smoke spreads laterally across the ceiling, forming a thin, high-velocity layer.
- *Layer growth*: As more smoke accumulates, the layer thickens downward.

Atrium Smoke Exhaust System Components

An effective atrium smoke exhaust system typically includes the following:

1. Smoke Exhaust Fans:
- Located at the top of the atrium, these remove the accumulated smoke layer.
- Sized based on the volumetric flow rate of smoke, which depends on the fire size and plume dynamics.

2. Makeup Air Openings:
- Provide replacement air to prevent negative pressure and ensure effective smoke removal*.
- Strategically located to avoid disrupting the plume or spreading smoke.

*when considering makeup air, it should be remembered that an extract system can only extract a volume equal to that of the incoming (makeup) air.

3. Ductwork:
- Transports smoke from the exhaust fans to the outside.
- Designed to minimize pressure losses and ensure sufficient flow rates.

4. Control Systems:
- Detect the fire and automatically activate the exhaust system.
- May include fire alarm systems integrated with temperature or smoke sensors.

5. Smoke Barriers:
- Prevent smoke from spreading to adjacent areas.
- Could include automatic curtains or fixed glass barriers.

Design Considerations with Axisymmetric Plumes

Plume Mass Flow Rate:
- The rate at which smoke and entrained air rise is a function of the fire size, height above the fire, and plume geometry.
- It is calculated using empirical formulas or simulations.

Plume Height:
- The exhaust system must be designed to operate effectively at the expected smoke layer interface height.

Smoke Layer Temperature:
- Determines the buoyancy and flow dynamics of the smoke.
- Affects material selection for the exhaust system and the sizing of fans.

Visibility and Toxicity:
- The system must maintain a clear layer for occupant evacuation, with visibility >10 m and CO concentrations within safe limits.

Axisymmetric Plume Equations

The following formulas describe the mass flow rate (m˙), velocity (u) and temperature (T) in an axisymmetric plume:

Mass Flow Rate:

$$m(z) = \pi k^2 p g^{0.5} Q^{1.5} z^{-0.5}$$

Where:

k : Entrainment coefficient

p : Air density

g : Gravitational acceleration

Q : Heat release rate of the fire

z : Height above the fire

Velocity and Temperature:

These decrease with increasing distance from the fire source due to entrainment and cooling.

Example of Application
- Consider a shopping mall atrium with a fire producing a 5 MW heat release rate:
- Calculate the plume's mass flow rate and temperature at specific heights using the axisymmetric equations.

- Ensure the design of the exhaust fans can handle the smoke volume at the predicted layer height.
- Ensure sufficient makeup air openings and barriers to contain the smoke.

Testing and Maintenance

To ensure reliability, atrium smoke exhaust systems must undergo regular testing and maintenance:
- Check fan performance and duct integrity.
- Verify sensor and control system functionality.
- Conduct smoke tests to simulate plume behaviour and system response.

This detailed design and operational approach ensure the safety of occupants and compliance with fire safety regulations, especially in large spaces with complex geometries like atriums.

Balcony Spill Plume Management

Atrium Smoke Exhaust System with a Balcony Spill Plume

A balcony spill plume is a specific type of smoke plume formed when smoke rises through an atrium from a fire located on a lower floor and spills over the edge of a balcony or mezzanine. The atrium smoke exhaust system for this scenario is designed to handle the unique flow patterns and characteristics of smoke as it transitions from a confined space (below the balcony) to an open space (the atrium). This design ensures effective smoke management, occupant safety, and compliance with fire safety regulations.

Balcony Spill Plume Characteristics

A balcony spill plume differs from an axisymmetric plume due to its geometry and flow dynamics:
- *Non-axisymmetric behaviour*: The plume flows horizontally and then, vertically, spilling over the edge of the balcony.
- *Smoke entrainment*: Ambient air is drawn into the plume as it flows over the edge, increasing its volume and diluting the smoke.
- *Complex flow pattern*: The interaction between the balcony edge and the atrium airflow creates a more turbulent flow compared to a free plume.

Smoke Layer Formation

In the event of a fire on a lower level:

- *Initial confinement*: Smoke accumulates below the balcony, forming a layer at the ceiling of the lower level.
- *Spillover*: The layer moves laterally and flows over the edge of the balcony into the atrium.
- *Vertical rise*: The spilled smoke rises into the atrium, forming a stratified layer near the atrium ceiling.

Atrium Smoke Exhaust System Components

To handle a balcony spill plume effectively, the atrium smoke exhaust system includes the following components:

1. **Smoke Exhaust Fans**:
   - Located at the top of the atrium to remove the accumulated smoke layer.
   - Sized to handle the increased volume of smoke due to entrainment at the balcony edge.
2. **Balcony Spill Edge Design**:
   - Designed to minimize turbulence and control smoke flow.
   - May include physical barriers or overhangs to direct the plume into the atrium.
3. **Makeup Air Openings**:
   - Strategically placed to provide replacement air and maintain pressure balance.
   - Avoid placement near the balcony edge to prevent disruption of the plume.
4. **Smoke Barriers**:
   - Curtains, partitions, or glazing to contain the smoke below the balcony and guide it towards the atrium.
5. **Control Systems**:
   - Sensors and alarms to detect smoke and activate the exhaust system.
   - Integrated with fire suppression systems for additional safety.

Design Considerations for Balcony Spill Plumes

The design of an atrium smoke exhaust system for balcony spill plumes must account for several factors:

1. **Spill Plume Mass Flow Rate:**
   - The rate at which smoke flows over the balcony edge is critical for sizing the exhaust system.
   - Empirical formulas or Computational Fluid Dynamics (CFD) models are used to predict flow rates

$$m(z) = Cpg^{0.5}Q^{0.667}z^{0.333}$$

Where:

C : Empirical coefficient depending on geometry and entrainment

p : Air density

g : Gravitational acceleration

Q : Heat release rate of the fire

z : Height above the balcony edge

2. **Entrainment Effects:**
   - Significant entrainment occurs at the balcony edge, increasing the smoke volume.
   - Exhaust system capacity must be sufficient to handle this increased flow.

3. **Smoke Layer Temperature and Visibility:**

The system must maintain a tenable environment in the atrium, with acceptable temperature and visibility levels for evacuation.

4. **Balcony Edge Dynamics:**

The shape and configuration of the balcony edge influence the plume behaviour. Smooth edges reduce turbulence, while sharp edges can increase smoke dispersion.

5. **Evacuation and Accessibility:**

Design should prioritize clear egress routes and avoid obstructing evacuation paths with smoke.

Example of System Design

Consider an atrium in a hotel with a fire on the second-floor balcony, producing a 3 MW heat release rate. Design steps include:
- *Estimate mass flow rate*: Use empirical formulas to calculate the smoke volume spilling over the balcony edge.

- *Exhaust fan sizing*: Select fans capable of removing the calculated smoke volume.
- *Makeup air placement*: Design openings to balance airflow without interfering with the plume.
- *CFD simulations*: Validate the system design by modelling plume behaviour and smoke movement.

Testing and Maintenance

Regular testing ensures system reliability:

- Perform smoke tests to simulate a balcony spill plume and verify system response.
- Inspect exhaust fans, ductwork, and smoke barriers for damage or obstructions.
- Test sensors and control systems for timely activation.

Challenges and Solutions

Some typical challenges and potential solutions are captured in the table below:

	Challenge	Solution
1	Entrainment increases smoke volume and dilutes visibility	Oversize exhaust fans and incorporate additional makeup air provisions
2	Complex flow patterns at the balcony edge	Conduct detailed CFD analysis to optimize balcony edge design and airflow management
3	Coordinating evacuation with smoke movement	Integrate the exhaust system with evacuation modelling and alarm systems

An atrium smoke exhaust system designed for balcony spill plumes requires careful consideration of the plume's unique characteristics, including its mass flow rate, entrainment, and turbulence at the balcony edge. A robust system ensures effective smoke control, providing a safe environment for evacuation and compliance with fire safety regulations.

Window Plume Management

Steady Atrium Smoke Exhaust System with a Window Plume

A window plume occurs when smoke is released from a fire through a vertical opening, such as a window, into an adjacent space like an atrium. This type of plume is characterized by strong entrainment and a unique flow pattern influenced by the geometry of the window and the heat release rate of the fire. A steady atrium smoke exhaust system for such scenarios is designed to maintain a clear smoke layer at the lower levels of the atrium, ensuring safe evacuation and reducing the risk of fire spread.

Window Plume Characteristics

A window plume is formed when smoke escapes through a vertical opening, interacting with the atrium environment. Key features include:

- **Vertical and Lateral Spread**: Smoke rises vertically and spreads laterally into the atrium, influenced by the window geometry.
- **Strong Entrainment**: The plume entrains a significant amount of ambient air, increasing its mass flow rate and reducing its temperature.
- **Buoyant Flow**: The upward movement is driven by the buoyancy of hot gases.
- **Intermittent or Steady Behaviour**: While window plumes are often unsteady, a steady exhaust system assumes constant fire conditions for design purposes.

Smoke Layer Formation

When a fire occurs in a compartment adjacent to an atrium:

- **Initial Escape**: Smoke accumulates in the compartment, flows through the window, and forms a plume in the atrium.
- **Vertical Rise**: The plume rises until it reaches the atrium ceiling.
- **Ceiling Jet**: The smoke spreads laterally along the ceiling, forming a stratified smoke layer.

Atrium Smoke Exhaust System Components

A steady smoke exhaust system for an atrium with a window plume includes:

Smoke Exhaust Fans

- Installed at the atrium's upper levels to extract the accumulated smoke layer.
- Sized based on the plume's mass flow rate and the expected heat release rate.

Makeup Air Provisions

- Openings or systems that provide replacement air to maintain pressure balance.
- Placed at low levels to avoid disrupting the window plume.

Window Geometry Optimization

- The size and position of the window influence the smoke flow pattern and entrainment rate.

- Larger windows increase the plume's mass flow rate.

Smoke Barriers

- Glass walls, curtains, or other barriers to contain smoke and guide its movement.

Control and Detection Systems

- Sensors to detect fire and smoke, triggering the exhaust system.
- Often integrated with fire suppression systems for enhanced safety.

Design Considerations

Designing an atrium smoke exhaust system for a window plume involves several critical factors:

Plume Mass Flow Rate

The mass flow rate of the window plume (m) increases with the heat release rate and window area. It is calculated as:

$$m(z) = C_w \cdot A_w \cdot p \cdot g^{0.5} \cdot Q^{0.333} \cdot z^{0.333}$$

Where:

C_w : Empirical coefficient based on window configuration.

A_w : Window area.

P : Air density.

g : Gravitational acceleration.

Q : Heat release rate of the fire.

z : Height above the window opening.

Entrainment and Visibility

- Entrainment of ambient air dilutes the smoke but increases its volume.
- The system must maintain tenable visibility and temperature levels for evacuation.

Smoke Layer Interface Height

- The exhaust system should maintain the smoke layer at a safe height above evacuation paths, typically 2.5 m or more.

Thermal Buoyancy

- The system must account for the buoyancy of the hot gases to optimize exhaust rates

Window Placement
- Higher window placement can reduce the risk of low-level smoke accumulation but may increase the smoke layer's temperature.

Example Design Scenario

Consider an atrium with a fire in an adjacent compartment, producing a 2 MW heat release rate, with a window area of 2 m². The design steps include:

- *Calculate Mass Flow Rate*: Use the window plume equation to estimate the flow rate at a specific height.
- *Determine Exhaust Fan Capacity*: Select fans capable of removing the calculated smoke volume, ensuring a steady-state condition.
- *Design Makeup Air Openings*: Provide low-level openings for air replenishment, avoiding interference with the window plume.
- *Perform CFD Simulations*: Validate the system's performance under steady conditions, ensuring effective smoke control.

Testing and Maintenance

Regular testing ensures system reliability and effectiveness:

- *Smoke Tests*: Simulate fire scenarios to observe plume behaviour and exhaust system response.
- *Fan Performance Checks*: Inspect and test fans for proper operation.
- *Sensor Functionality*: Verify the accuracy and responsiveness of smoke and fire detection systems.
- *Ductwork Inspection*: Ensure ducts are free of obstructions and leaks.

Challenges and Solutions

Some key challenges and solutions in design are summarised in the following table:

	Challenge	Solution
1	High entrainment increases smoke volume	Oversize the exhaust fans to accommodate entrained air
2	Uneven smoke layer due to window geometry	Optimize window design and use barriers to guide the plume
3	Disrupted plume flow due to cross-ventilation	Carefully position makeup air openings to avoid interference

A steady atrium smoke exhaust system designed for a window plume must address the unique characteristics of smoke flow through vertical openings. By optimizing fan capacity, window geometry, and airflow management, the system ensures effective smoke removal, occupant safety, and compliance with fire safety regulations. Advanced simulations and regular maintenance further enhance system reliability in real-world fire scenarios.

Communicating Space Smoke Plume Management

Airflow to Control Smoke Flow from a Communicating Space

In fire safety engineering, managing airflow to control smoke flow from a communicating space is a critical strategy to ensure safe evacuation, prevent smoke spread, and maintain tenable conditions in adjacent areas. Communicating spaces, such as open-plan atriums connected to corridors, lobbies, or stairwells, present unique challenges due to their shared volumes and open pathways for smoke movement.

Challenges of Smoke Flow in Communicating Spaces

Smoke flow in communicating spaces is influenced by several factors:

Buoyancy: Hot smoke is buoyant and rises, creating stratification and potential layering.

Pressure Differences: Driven by thermal expansion of smoke and wind or mechanical ventilation systems.

Geometry: Openings, walls, and ceiling heights impact how smoke spreads.

Stack Effect: In tall buildings, the difference in temperature between indoor and outdoor air can cause vertical smoke movement.

Entrainment: Airflow into the smoke layer dilutes and cools the smoke, affecting its movement.

Objectives of Airflow Control

Effective airflow control aims to:

Contain Smoke: Prevent smoke from spreading to adjacent spaces or evacuation routes.

Maintain Tenable Conditions: Ensure clear visibility, breathable air, and safe temperatures for occupants.

Facilitate Smoke Exhaust: Direct smoke toward exhaust outlets for removal.

Methods for Airflow Control in Communicating Spaces

Several strategies can be employed to control airflow and smoke movement:

1. **Mechanical Smoke Control Systems**

   Smoke Exhaust Systems:
   - Extract smoke from the communicating space using exhaust fans located near the ceiling.
   - Creates a low-pressure zone to encourage smoke to rise and stay away from evacuation paths.

   Pressurization Systems:
   - Increase air pressure in adjacent spaces (e.g., stairwells, lobbies) to prevent smoke from entering.
   - Typically, fans supply air at a higher pressure than the fire compartment.

2. **Compartmentalized Ventilation:**
   - Isolates airflow between communicating spaces using fire-rated barriers and dampers.

3. **Passive Airflow Control**

   Smoke Barriers:
   - Physical barriers like fire curtains, walls, or glazing prevent smoke spread into adjacent areas.
   - Used in atriums and open plan designs to compartmentalize smoke.

   Natural Ventilation:
   - Utilize openings like windows or vents to direct smoke upward and out of the space. Often combined with thermal buoyancy effects.

Airflow Management Techniques

Airflow Across Openings:
- Maintain sufficient airflow across openings to block smoke movement into adjacent areas. Airflow velocity must be high enough to counteract smoke pressure differences.
- Minimum airflow velocity is calculated as:

$$v = \sqrt{\frac{2\Delta p}{p}}$$

Where:

v : Airflow velocity (m/s)

Δp : Pressure difference across the opening (Pa)

p : Air density (kg/m³)

Makeup Air Supply
- Provide replacement air to prevent negative pressure in the communicating space and stabilize smoke movement.
- Makeup air should not disturb the smoke plume or reduce exhaust efficiency.

Flow Direction Control
- Use dampers or adjustable vents to guide smoke flow toward exhaust points and away from critical spaces.

Design Considerations

When designing airflow control systems for communicating spaces, the following factors must be considered:

Fire Size and Heat Release Rate (HRR)
- Larger fires produce more smoke, requiring higher airflow rates for effective control.

Smoke Layer Thickness
- Ensure adequate distance between the smoke layer and occupied zones or evacuation paths.

Geometry of the Space
- The height, volume, and configuration of openings affect airflow patterns and smoke behaviour.

Environmental Conditions
- Wind, temperature gradients, and stack effects can influence airflow and smoke spread.

Regulatory Compliance
- Adhere to local fire codes and standards, such as NFPA 92 or EN 12101, for smoke control design.

Example Scenario

Scenario:

An atrium with a connecting corridor experiences a fire on the atrium floor.

Problem:

Smoke begins to rise into the atrium and spread toward the corridor.

Solution:
- *Exhaust Smoke from the Atrium*:

Use smoke exhaust fans to remove smoke from the atrium ceiling, maintaining a clear layer.

- *Pressurize the Corridor:*

  Install fans to pressurize the corridor, creating an airflow barrier that prevents smoke entry.

- *Provide Makeup Air:*

  Add low-level air inlets to ensure stable smoke flow toward the exhaust points without disrupting the plume.

Testing and Maintenance

- To ensure system reliability:

Conduct Smoke Tests

- Simulate fire conditions to verify airflow patterns and smoke containment.

Inspect Fans and Ducts

- Check for blockages, leaks, or mechanical failures.

Monitor Airflow Sensors

- Regularly calibrate sensors for accurate pressure and velocity measurements.

Airflow control in communicating spaces is vital to managing smoke during a fire. By employing a combination of mechanical and passive systems, optimizing airflow velocity and direction, and considering the space's geometry and fire dynamics, effective smoke containment and removal can be achieved. Regular testing and maintenance further ensure that these systems perform as intended during critical events.

Management of Smoke Flow From Smoke Layer

Controlling Smoke Flow from the Smoke Layer

In fire safety engineering, controlling smoke flow from a smoke layer is essential to maintain tenable conditions for building occupants, ensure effective evacuation, and facilitate fire suppression efforts. The smoke layer, formed when hot gases from a fire rise and stratify near the ceiling, is a critical aspect of smoke management in enclosed or semi-enclosed spaces like atriums, tunnels, or large halls.

Characteristics of the Smoke Layer

The smoke layer is influenced by:

Buoyancy

- Hot gases rise due to lower density, forming a stratified layer near the ceiling.

Temperature Gradient

- The layer's temperature decreases as it entrains cooler ambient air.

Thickness
- Determined by the fire's heat release rate (HRR), room geometry, and ventilation conditions.

Smoke Flow
- The layer spreads laterally across the ceiling, forming a ceiling jet before descending as it thickens.

Objectives of Smoke Layer Control

Effective smoke layer control aims to:

Prevent Smoke Spread
- Limit smoke movement to adjacent areas or evacuation routes.

Maintain Visibility
- Ensure visibility for occupants and first responders, typically >10 m in escape routes.

Ensure Safe Temperatures
- Keep temperatures below tenable limits (e.g., 60°C for evacuation routes).

Facilitate Exhaust and Ventilation
- Enable efficient smoke removal to delay smoke layer descent.

Methods for Controlling Smoke Flow from the Smoke Layer

1. Smoke Exhaust Systems

Mechanical Exhaust Fans
- Remove smoke from the upper levels of a space, reducing the layer's thickness and preventing it from descending.
- Fans are sized based on the volumetric flow rate of the smoke layer:

$$Q = m / p$$

Where:

Q: Volumetric flow rate (m³/s)

m: Mass flow rate of the smoke plume (kg/s)

p: Air density (kg/m³)

Natural Ventilation
- Utilize openings like windows, vents, or skylights to allow smoke to escape.

- Relies on buoyancy-driven flow and is more effective in spaces with large vertical openings.

2. **Smoke Reservoirs**
   - Contain the smoke layer within designated areas near the ceiling to facilitate exhaust.
   - Reduce the overall volume of smoke to be managed by the system.

3. **Makeup Air Supply**

   Low-Level Openings
   - Introduce cool, clean air at lower levels to replace the exhausted smoke and maintain pressure balance.
   - Prevents reverse flow into the exhaust system.

   Directional Airflows

   Use fans or vents to guide makeup air away from the smoke plume and toward the exhaust points.

4. **Active Control Systems**

   Dynamic Exhaust Rates
   - Adjust fan speeds based on real-time smoke layer conditions (e.g., thickness, temperature).

   Air Curtains
   - Create airflow barriers that direct smoke toward exhaust outlets while protecting adjacent spaces.

5. **Passive Measures**

   Smoke Barriers
   - Fire-resistant walls, curtains, or glazing to contain the smoke layer in specific zones.
   - Prevent lateral spread into unaffected areas.

   Ceiling Jet Modifiers
   - Physical structures or deflectors to disrupt or channel the ceiling jet's lateral flow.

Design Considerations for Smoke Layer Control

1. **Heat Release Rate (HRR)**
   - The HRR determines the smoke generation rate and influences the required exhaust system capacity.
   - Larger fires produce more smoke, requiring higher exhaust rates.

2. **Room Geometry**

   Ceiling Height
   - Taller spaces allow for greater smoke layer stratification and delay descent.

   Volume
   - Larger volumes require higher exhaust capacities or additional smoke reservoirs.

   Smoke Layer Thickness
   - A minimum safe distance (typically 2.5 m or more) must be maintained between the smoke layer's base and the occupied zone

   Environmental Conditions
   - Wind pressure, stack effects, and ambient temperature can affect the smoke layer's behaviour.

   Compliance with Standards
   - Systems must meet standards like NFPA 92, EN 12101, or ASHRAE guidelines for smoke control.

Example of Smoke Layer Control

Scenario:

A fire occurs in a large atrium, generating a smoke layer 1 m thick at the ceiling.

Problem:

The smoke layer begins to descend, threatening evacuation paths.

Solution:

- *Install Exhaust Fans*:

Extract smoke at a volumetric flow rate calculated for the fire's HRR and plume behaviour.

- *Add Makeup Air Openings*:

Provide low-level air inlets to replace exhausted smoke and stabilize the pressure.

- *Deploy Smoke Barriers*:

Use curtains to compartmentalize the smoke layer near the ceiling, preventing lateral spread.

Testing and Maintenance

1. **Regular Inspections:**

   Inspect fans, ducts, and smoke barriers for damage or blockages.

2. **Smoke Tests:**
Simulate fire scenarios to observe smoke flow and layer control effectiveness.
3. **Sensor Calibration:**
Ensure temperature, velocity, and smoke detectors are functioning accurately.

Challenges and Solutions

Select challenges and possible solutions are listed in the table below.

	Challenge	Solution
1	Smoke Layer Descent	Increase exhaust capacity or deploy additional makeup air sources
2	Uneven Smoke Distribution	Use smoke barriers or ceiling jet modifiers to guide smoke toward exhaust points
3	Cross-Ventilation Interference	Optimize inlet and outlet placement to prevent conflicting airflow patterns

Controlling smoke flow from a smoke layer is vital for maintaining safe evacuation conditions and ensuring the effectiveness of fire safety measures. By combining mechanical exhaust systems, makeup air strategies, and passive containment measures, engineers can manage smoke layers efficiently in various building configurations. Regular testing and adherence to fire safety standards ensure that these systems perform as intended during emergencies.

Management of Airflow To Control Smoke Flow From Plume

Airflow to Control Smoke Flow from the Plume

Controlling airflow to manage smoke flow from the plume is a critical aspect of fire safety engineering. The plume represents the buoyant flow of hot gases and smoke rising from the fire. Effective management ensures that the plume is directed toward smoke exhaust systems and does not compromise evacuation routes or adjacent spaces.

Understanding Smoke Plume Dynamics

The behaviour of the smoke plume is influenced by:

Buoyancy
The heat from the fire reduces the density of the smoke, causing it to rise.

Entrainment

As the plume rises, it entrains surrounding air, increasing its volume and decreasing its temperature.

Flow Patterns

The plume spreads vertically and laterally, forming a ceiling jet upon reaching the roof or other horizontal obstructions.

Objectives of Airflow Control

The primary goals of controlling airflow around the plume are:

Directing Smoke to Exhaust Systems

- Guide the plume toward exhaust points to ensure efficient removal.

Preventing Smoke Spread

- Avoid the lateral spread of smoke into occupied or escape areas.

Maintaining Visibility and Safe Conditions

- Ensure smoke does not obstruct visibility or create untenable conditions in evacuation paths.

Methods of Airflow Control

Mechanical Smoke Control Systems

Smoke Exhaust Fans

- Installed at the top of the space to remove smoke and prevent its accumulation near the ceiling.
- Sizing is based on the plume's mass flow rate:

$$m(z) = C \cdot p \cdot g^{0.5} \cdot Q^{0.0667} \cdot z^{0.333}$$

Where:

C : Empirical coefficient.

p : Air density.

g : Gravitational acceleration.

Q : Heat release rate of the fire.

z : Height above the fire.

Induced Airflows
- Use mechanical systems to create airflow patterns that guide the plume directly toward exhaust systems.
- Velocity near the plume must avoid disrupting its buoyant rise.

Compartmentalized Ventilation
- Divide the space into zones, with dedicated ventilation systems for each zone to isolate and control the plume.

Passive Airflow Strategies

Natural Ventilation
- Utilize buoyancy-driven flow through openings or vents to direct smoke upward and outward
- Requires carefully placed high-level vents to maximize smoke removal.

Smoke Barriers
- Install physical barriers (e.g., curtains, partitions) to contain the plume and prevent lateral spread.

Ceiling Features
- Use ceiling configurations or deflectors to channel the ceiling jet and guide smoke toward exhaust outlets.

Airflow Management

Airflow Velocity Control
- Ensure airflow velocity is sufficient to influence smoke movement without causing turbulence or disrupting the plume's buoyancy.
- For cross flow systems, maintain an airflow velocity greater than the plume rise velocity to contain lateral smoke spread.

Makeup Air Supply
- Introduce replacement air at low levels to stabilize pressure and ensure the plume flows upward.
- Makeup air must not interfere with the rising plume or entrain smoke into occupied areas.

Active Control Systems
Real-Time Adjustments
Sensors
Use sensors to monitor smoke behaviour (e.g., temperature, velocity) and adjust fan speeds or vent openings dynamically.

Air Curtains
Create high-velocity airflows at boundaries to contain the plume and direct smoke toward exhaust systems.

Design Considerations
Fire Characteristics:
Heat Release Rate (HRR)
- Determines the initial velocity and buoyancy of the plume.

Smoke Composition
- Heavier or cooler smoke may require additional airflow control measures.

Geometry of the Space:
Ceiling Height
- Higher ceilings allow for greater plume rise before interacting with the roof.

Openings
- Placement of vents or barriers affects airflow patterns and smoke movement.

Environmental Conditions
- Wind, temperature gradients, and stack effects can influence plume behaviour and airflow.

Standards Compliance
- Adhere to guidelines like NFPA 92 or EN 12101 for plume control and exhaust system design.

Example Application
Scenario:
A fire in a shopping mall generates a smoke plume with an HRR of 3 MW.

Problem:
The smoke plume is spreading laterally and descending, threatening evacuation paths.

Solution:

Install Exhaust Fans:
- Calculate the mass flow rate and size fans accordingly.

Deploy Makeup Air Openings:
- Provide low-level air inlets to stabilize the pressure.

Use Smoke Barriers:
- Guide the plume directly toward the exhaust outlets.

Testing and Maintenance

Smoke Tests
- Simulate fire scenarios to validate plume control strategies.

Fan and Sensor Inspections
- Regularly inspect and test smoke exhaust fans and monitoring devices.

System Calibration
- Ensure real-time control systems respond effectively to changes in plume dynamics.

Challenges and Solutions

Challenges and suggested solutions in this area are given below.

	Challenge	Solution
1	Turbulence Disrupting the Plume	Optimize airflow velocity and avoid high-velocity makeup air near the plume
2	Incomplete Smoke Removal	Add additional exhaust points or increase fan capacity
3	Cross-Ventilation Effects	Ensure proper placement of vents and barriers to control airflow direction

Airflow control for smoke plumes requires a combination of mechanical, passive, and active strategies to ensure effective smoke removal and containment. By considering plume dynamics, space geometry, and environmental conditions, engineers can design systems that protect occupants and maintain compliance with fire safety standards. Regular testing and maintenance are essential to ensure system reliability during fire emergencies.

Chapter 6

Protecting Lives and Property

Fire engineering plays a crucial role in safeguarding human life by minimising casualties and injuries, as also limiting damage to property, during fires. The processes and practices involved can be broadly categorized into:

- Pro-active measures to prevent fires and minimize damage to life and property
- Post incident analysis

Pro-Active Measures To Prevent Fires

The adage *'prevention is better than cure'* applies equally well to the threat of fires. It is imperative that robust preventive measures are taken to minimize the risk of fire and consequent losses. Such measures ensure that the desired outcomes are achieved with less energy, cost, and also greater peace of mind. If, despite these measures, an incident does occur, the measures help contain and limit the damage to life and property. The steps that can be taken to achieve this can be divided broadly into:

- Design, including:
    - structure
    - usage of appropriate fire-resistant materials
    - egress and evacuation planning for occupants
    - application of technology
    - leveraging innovation and
    - environmental considerations
- Active Fire Protection Systems
- Education & Emergency Preparedness
- Emergency Response Planning
- Compliance and Regulatory Bodies
- Fire Safety Codes and Regulations

Design

Structure & Materials of construction

Sensitivity to the need for fire protection needs to begin from the stage of planning the construction of a building. The very structure of the building can be designed in such a manner that the risk of fires is minimized. Utilising fire-resistant materials in construction helps slow the spread of fire, providing occupants with more time to escape. Similarly, designing buildings with fire compartments can contain fires within specific areas, limiting damage and protecting escape routes.

Egress and Evacuation Planning

Clear Exit Routes: Designing buildings with clearly marked and unobstructed exit routes ensures that occupants can evacuate themselves quickly and safely.

Emergency Lighting: Installing emergency lighting helps guide people to exits during smoke-filled or dark environments.

Technology

Technologies like *Artificial Intelligence (AI), The Internet of Things (IoT), and Fire modelling software* are revolutionizing the field of fire engineering by enhancing predictive capabilities, improving safety measures, and optimizing emergency responses.

AI algorithms can analyse vast amounts of data to identify patterns and potential fire risks, enabling proactive risk management and faster decision-making. For example, AI can predict fire behaviour based on environmental conditions, allowing engineers to design more resilient structures.

IoT facilitates real-time monitoring of fire safety systems through interconnected devices. Smart sensors can detect smoke, heat, or gas and automatically alert emergency services, reducing response times and improving outcomes during a fire incident. IoT devices can also gather data on building occupancy and conditions, providing valuable insights for fire safety planning and management.

Fire modelling software allows engineers to simulate fire scenarios and assess the effectiveness of various safety measures. By creating detailed models of fire spread and smoke movement, engineers can better understand potential hazards and refine designs to enhance safety. This software enables more informed decision-making during the design phase and aids in developing comprehensive evacuation plans.

Together, these technologies are transforming fire engineering by promoting a proactive approach to fire safety, improving preparedness, and ultimately saving lives and property.

Innovation

Ongoing research and innovation are essential in advancing fire engineering and improving fire safety measures. As new materials, technologies, and construction practices emerge, continuous investigation helps identify potential fire hazards and develop more effective prevention and mitigation strategies. Research enables the understanding of fire behaviour, allowing engineers to design structures that can better withstand flames and smoke, thereby protecting occupants and minimizing property damage.

Innovation also drives the development of smarter fire detection and suppression systems, enhancing response times and reducing the risk of catastrophic events. Furthermore, a commitment to research fosters a culture of knowledge sharing and collaboration within the industry, ensuring that the latest findings are disseminated and implemented effectively. Ultimately, prioritizing ongoing research and innovation is crucial for adapting to evolving fire risks and enhancing overall safety in built environments.

Environment

It is important to examine how fires can have detrimental effects on the environment, including air and water pollution. Integrating sustainability into fire engineering is crucial for fostering a safer and more environmentally responsible built environment. By utilizing environmentally friendly materials and technologies, fire engineers can reduce the ecological impact of construction and renovation projects. Sustainable practices, such as selecting low-emission materials and implementing energy-efficient fire suppression systems, not only enhance the resilience of structures against fire hazards but also contribute to overall environmental health. This holistic approach helps mitigate climate change effects and promotes resource conservation, ensuring that fire safety measures align with broader sustainability goals. Ultimately, prioritizing sustainability in fire engineering supports the development of safer communities while protecting our planet for future generations.

Active Fire Protection Systems

The multiple systems available to prevent and control fires, as also to help in fire management, are listed below. Given their importance, these will be discussed in more detail in a subsequent chapter.

- **Fire Detection and Alarm Systems**

  Automatically detect fire or smoke and alert building occupants and emergency services allowing prompt evacuation.

- **Sprinkler Systems**

  They can automatically detect fires and discharge water to suppress or extinguish flames, thus halting spread and significantly reducing the potential for injury.

- **Water Mist Systems**

  Use fine water sprays to cool flames and surrounding gases, reducing fire intensity and spread.

- **Gaseous Fire Suppression Systems**

  Release inert or chemical gases to suppress fires without damaging sensitive equipment.

- **Foam Suppression Systems**

  Deploy foam to smother fires involving flammable liquids by cutting off oxygen supply.

- **Dry Chemical Suppression Systems**

  Discharge dry chemical agents to interrupt the chemical reaction of a fire.

- **Wet Chemical Suppression Systems**

  Typically used in commercial kitchens to suppress grease fires and prevent re-ignition.

- **Smoke Control Systems**

  Extract or contain smoke to maintain escape routes and improve visibility for evacuation and firefighting.

- **Fire Curtain Systems**

  Automatically deploy fire-resistant barriers to compartmentalise fire and smoke in open-plan areas.

- **Fire Extinguishers**

  Portable devices used to manually suppress small fires in the early stages.

- **Fire Hose Reels**

  Provide a steady supply of water for occupants or firefighters to control and extinguish fires.

- **Emergency Voice Communication Systems (EVCS)**

  Enable fire wardens and emergency services to communicate during a fire event.

- **Automatic Fire Doors**

   Close automatically when a fire is detected to prevent spread between compartments.

- **Firefighter Lift Systems**

   Elevators designed with fire-resistant features to assist firefighter access during emergencies.

Education, Training & Emergency Preparedness

Training is crucial for both individuals and organizations in minimizing fire risks and enhancing overall safety. For individuals, training helps raise awareness about potential fire hazards and teaches essential skills such as how to use fire extinguishers, identify escape routes, and perform basic first aid. By understanding the nature of fire and the importance of preparedness, individuals are more likely to act quickly and effectively in an emergency, potentially saving lives and reducing property damage.

For organizations, comprehensive fire preparedness training is vital in ensuring that all employees are equipped to respond appropriately to fire incidents. This includes conducting regular fire drills, implementing emergency evacuation plans, and familiarizing staff with fire safety equipment. Such training fosters a culture of safety within the workplace, encouraging teamwork and clear communication during emergencies.

Moreover, preparedness training can help organizations comply with safety regulations and standards, thereby minimizing legal liabilities. By investing in training programs, both individuals and organizations not only enhance their readiness to face fire emergencies but also create safer environments that can significantly reduce the likelihood of fire incidents occurring in the first place. Overall, preparedness training is a proactive strategy that empowers communities and workplaces to mitigate fire risks effectively.

Role of Fire Engineers

Fire engineers play a vital role in society by ensuring the safety and protection of people, property, and the environment from fire hazards. Their expertise encompasses a wide range of responsibilities, including the design and evaluation of fire protection systems, conducting risk assessments, and developing fire safety codes and standards. One of their key roles is to collaborate with architects and builders to incorporate fire safety measures into building designs, ensuring that structures are resilient to fire risks. They also conduct fire investigations to determine the causes of incidents, which helps inform future safety practices and regulations.

In addition to their technical responsibilities, fire engineers serve as educators and advocates, raising public awareness about fire safety and prevention. They often engage with communities, providing training and resources to empower individuals to take proactive measures against fire risks.

Providing fire drills and safety training educates occupants on emergency procedures, helping them respond effectively during a fire. Public education campaigns about fire safety can raise awareness and prepare individuals for emergencies.

Moreover, fire engineers are integral in emergency planning and response efforts, working alongside fire departments and emergency services to develop effective strategies for managing fire incidents. Their work contributes to creating safer environments, ultimately enhancing the overall quality of life in communities. Through their multifaceted roles, fire engineers help build a culture of safety and resilience that benefits society as a whole.

Fire engineers play a vital role in educating communities about fire safety and prevention measures, serving as crucial advocates for public awareness and preparedness. Their expertise allows them to translate complex fire safety concepts into accessible information for diverse audiences, from schools to local organizations. By conducting workshops, seminars, and outreach programs, fire engineers can teach residents about fire hazards, safe practices, and the importance of having emergency plans in place.

They also collaborate with local fire departments to create educational materials and campaigns that highlight the significance of smoke alarms, fire extinguishers, and evacuation routes. Additionally, fire engineers contribute to community planning by advising on building codes and safety regulations, ensuring that new developments prioritize fire safety from the outset.

Through these efforts, fire engineers empower individuals and communities to take proactive measures, ultimately fostering a culture of safety and reducing the risk of fire-related incidents. Their educational initiatives not only enhance public knowledge but also cultivate resilience, equipping communities to respond effectively in emergencies.

Emergency Response Planning

Collaborating with the local Fire & Rescue Service to develop emergency response plans can significantly enhance overall preparedness in several ways:

- Local fire departments have specialised knowledge and experience in handling emergencies. Collaborating with them allows organisations to incorporate best practices and insights into their emergency response plans.

- Fire departments can provide valuable input on specific risks associated with the local area, helping to develop tailored strategies that address unique hazards, such as high-rise buildings, industrial zones, or rural locations.
- Joint training exercises and drills with fire personnel can improve readiness among staff and emergency responders. These activities help identify gaps in plans and provide practical experience in responding to real-life scenarios.
- Collaborating can lead to better resource allocation and utilisation. Fire departments can advise on the necessary equipment and personnel needed during an emergency, ensuring that both the organisation and fire services are adequately prepared.
- Establishing strong communication links between organizations and local fire departments ensures that everyone is on the same page during an emergency. This clarity can reduce response times and improve coordination.
- Joint efforts in public education campaigns about fire safety and emergency preparedness can enhance community resilience. Fire departments can help disseminate critical information and resources to the public.
- Regular reviews and updates of emergency response plans, in collaboration with fire services, ensure that they remain effective and reflect any changes in local risks, building codes, or fire safety regulations.
- Collaborating with fire departments fosters a more integrated approach to emergency management, aligning responses with other local services like police and medical responders.
- Engaging with fire services can help ensure that emergency response plans comply with relevant fire safety regulations and standards, reducing legal risks.
- Strong partnerships with local fire departments can facilitate a community-wide culture of preparedness, leading to a more resilient environment that can effectively respond to emergencies.

By leveraging the expertise and resources of local fire departments, organizations can significantly improve their emergency response plans, ultimately enhancing safety for employees, visitors, and the surrounding community. Strategically placing fire-fighting resources and personnel ensures a rapid response to fires, reducing potential casualties.

Compliance & Regulatory Bodies

Compliance with Regulations

Adhering to local and international building codes and standards is vital for fire engineers to ensure safety and prevent disasters. These codes provide a framework for best practices in design, construction, and maintenance, specifically addressing fire prevention and protection measures. Compliance helps to establish minimum safety requirements that safeguard lives, property, and the environment by minimizing fire risks and enhancing emergency response capabilities. Additionally, following these regulations fosters consistency and accountability within the industry, ensuring that all stakeholders understand their responsibilities. By prioritizing adherence to established standards, fire engineers can effectively mitigate potential hazards, promote resilience in the built environment, and ultimately contribute to a culture of safety that benefits communities as a whole.

The Role of Regulatory Bodies

Organizations like the UK fire safety organizations play a crucial role in establishing safety standards by conducting research, developing guidelines, and promoting best practices within the fire safety sector. These organizations, such as the National Fire Chiefs Council (NFCC) and the Fire Industry Association (FIA), collaborate with government bodies, industry experts, and stakeholders to identify emerging risks and formulate comprehensive safety policies. They provide essential training and resources, ensuring that fire engineers and professionals are well-equipped with the latest knowledge and techniques. Additionally, these organizations often engage in advocacy efforts to influence legislation and public awareness regarding fire safety. By fostering a culture of continuous improvement and knowledge sharing, they help ensure that safety standards evolve in response to new challenges, ultimately enhancing the protection of lives and property across the UK.

Fire Safety Codes & Regulations

Building Codes Compliance: Adhering to local and national fire safety codes ensures that buildings are constructed with necessary safety features. There are numerous fire safety building codes in the UK; while the following list is not exhaustive it does contain the most used:

- **Building Regulations 2010**: Part B of the Building Regulations covers fire safety in England and Wales. It includes provisions for fire safety design in buildings.

- **The Regulatory Reform (Fire Safety) Order 2005**: This legislation applies to non-domestic premises and outlines the responsibilities of employers and building owners regarding fire risk assessments and fire safety management.
- **The Fire Safety Act 2021**: This act clarifies the scope of the Fire Safety Order and reinforces fire safety responsibilities, particularly in multi-occupied residential buildings.
- **National Fire Safety Guidance (various documents)**: The government publishes guidance documents that provide detailed advice on complying with fire safety laws, including:
  - **Approved Document B:** Fire safety, which provides guidance on meeting the requirements of the Building Regulations.
  - **LGA Fire Safety Guidance:** Guidance from the Local Government Association, which includes best practices for fire safety in housing.
- **British Standards (BS)**: Specific British Standards provide detailed guidelines on fire safety measures, such as:
  - **BS 9999**: Code of practice for fire safety in the design, management, and use of non-domestic buildings.
  - **BS 9991**: Code of practice for fire safety in the design, management, and use of residential buildings.
  - **BS 7974: 2019**: This provides a framework for the application of fire safety engineering principles to the design of buildings. This standard provides a framework for fire safety engineering and outlines methodologies for assessing fire safety performance in buildings. It includes guidance on the development of fire safety strategies, fire modelling, and the assessment of fire risks, helping designers and engineers to create safer buildings.
    - Additionally, the standard emphasises the importance of a holistic approach to fire safety, integrating aspects such as fire detection, suppression, and evacuation strategies into the design process.
    - For more comprehensive fire safety engineering practices, it is also beneficial to look at related standards, such as those in the **BS EN 1991-1-2** (Eurocode 1: Actions on structures – Part 1-2: General actions – Actions on structures exposed to fire) and **BS EN 1363** series (Fire resistance testing).
- **Local Authority Regulations**: Local councils may have additional fire safety regulations and requirements that need to be followed.

- **Health and Safety at Work Act 1974**: While not exclusively about fire safety, this act requires employers to ensure the safety of their employees, which includes managing fire risks.
- **Fire Safety Risk Assessment Guides**: These are published by various organisations, including the government and fire safety bodies, to help businesses and property owners assess and mitigate fire risks.
- **Industry-Specific Codes**: Certain industries may have additional fire safety regulations, such as those governing the construction of high-rise buildings or care homes.

It is important to consult the latest versions of these regulations and standards, as they can be updated. For specific projects, consulting with fire safety professionals or building inspectors is often recommended to ensure compliance.

Post Incident Analysis

Analysing past fires helps identify weaknesses in fire protection strategies and informs improvements for future designs and practices. Implementing lessons learned from fire incidents into engineering practices can enhance safety in future developments. This works like a typical feedback loop, learning from mistakes and adapting to changing circumstances, leading, ultimately, to continuous improvement.

Effective fire engineering practices are integral to minimizing casualties and injuries during a fire. By focusing on building design, active and passive fire protection measures, education, and compliance with safety regulations, fire engineering not only enhances safety but also fosters a culture of preparedness. This holistic approach significantly increases the chances of survival and reduces the impact of fire emergencies on human life and property.

Chapter 7

Fire Risk Management

Introduction

Fire risk management is a critical aspect of fire safety engineering, ensuring that buildings and environments are assessed, designed, and managed to mitigate fire hazards. Fire risk assessors and fire engineers play distinct yet complementary roles in fire risk management, from identifying potential risks to implementing engineered fire safety solutions.

Fire Risk Management Goals

The primary objectives of fire risk management include:

Life Safety: Protecting occupants from fire-related hazards through effective fire safety planning.

Asset Protection: Minimizing property damage through active and passive fire protection strategies.

Business Continuity: Ensuring that fire incidents do not disrupt business operations and productivity.

Environmental Protection: Reducing fire-related pollution, such as toxic emissions from burning materials.

Emerging Challenges in Fire Risk Management

- **Lithium-ion Battery Fires:** Increased use in vehicles, solar panel installations and consumer electronics presents unique risks.
- **High-Rise Fire Safety:** Urbanization has led to taller buildings requiring advanced evacuation strategies.
- **Sustainability Considerations:** Green materials and energy-efficient designs must align with fire safety principles.

Role of Fire Risk Assessors

Fire risk assessors focus on evaluating the potential for fire hazards in a given environment and ensuring compliance with fire safety regulations. Their responsibilities include:

- Conducting fire risk assessments per legal requirements (e.g., Regulatory Reform (Fire Safety) Order 2005 in the UK).
- Identifying ignition sources, fuel loads, and potential fire spread pathways.
- Evaluating the adequacy of fire detection, suppression, and evacuation systems.
- Assessing the vulnerability of occupants, including those with disabilities.
- Recommending risk mitigation measures such as fire compartmentation, emergency lighting, and exit route improvements.
- Documenting findings and advising building owners and responsible persons on fire safety compliance.

Legal Liabilities of Fire Risk Assessors

Failure to comply with fire safety regulations can result in legal consequences such as fines, penalties, or imprisonment. It is crucial that risk assessors document assessments thoroughly and ensure regular updates in compliance with evolving fire safety laws.

Technological Advances in Fire Risk Assessments

- **IoT-Based Fire Monitoring:** Sensors and smart alarms provide real-time fire risk monitoring.
- **AI-Powered Risk Analysis:** Machine learning models predict fire hazards based on data trends.
- **Drones for Fire Risk Inspections:** Drones are used to inspect hard-to-reach areas in high-rise buildings.

Fire Risk Assessment for Heritage Buildings

- **Challenges:** Preserving historical architecture while integrating modern fire safety measures.
- **Solutions:** Implementing discreet fire suppression systems like water mist technology and heat-resistant fire barriers.

Role of Fire Engineers

Fire engineers apply scientific and engineering principles to analyse fire behaviour and design fire safety solutions. Their expertise includes:

- Developing fire strategies aligned with building codes and performance-based design.

- Using computational models (e.g., CFD fire modelling, evacuation simulations) to predict fire growth and occupant egress.
- Designing active fire protection systems such as sprinklers, smoke control, and suppression systems.
- Evaluating structural fire resistance and material performance under fire conditions.
- Collaborating with architects and regulatory bodies to integrate fire safety into building design.
- Conducting forensic investigations to analyse fire incidents and recommend design improvements.

Innovative Fire Engineering Solutions

- **AI-Based Fire Detection Systems:** Algorithms that detect early-stage fires from sensor data.
- **Hybrid Gas Suppression Systems:** Using advanced gas combinations to improve fire suppression efficiency.
- **Self-Healing Fire-Resistant Materials:** Fireproof coatings that regenerate after exposure to fire.

Sustainability in Fire Engineering

- **Eco-friendly Fire Retardants:** Developing biodegradable fire suppression chemicals.
- **Green Building Certifications:** Fire-safe materials that comply with LEED and BREEAM standards.

Fire Risk Assessment Methodology

Fire risk assessors follow a systematic approach to evaluating fire hazards and ensuring compliance:

- **Identification of Hazards** – Reviewing ignition sources, fuel sources, and potential fire spread mechanisms.
- **Assessment of People at Risk** – Determining the number and type of occupants, including those requiring assistance.
- **Evaluation of Existing Fire Safety Measures** – Analysing fire detection, suppression, compartmentation, and means of escape.
- **Risk Analysis and Recommendations** – Identifying high-risk areas and suggesting corrective actions.

- **Documentation and Review** – Maintaining records, issuing reports, and ensuring periodic reassessment.

Emergency Preparedness & Human Behaviour
- Understanding how individuals respond under fire conditions helps optimize evacuation planning.
- Training drills and fire safety education improve response efficiency.

Fire Engineering Design Principles

Fire engineers use analytical and simulation-based approaches to design effective fire protection strategies:

- **Fire Load Analysis:** Calculating potential heat release rates based on building materials and fuel sources.
- **Smoke Management:** Designing ventilation and smoke extraction systems to maintain tenable conditions.
- **Evacuation Modelling:** Simulating occupant movement using software like Pathfinder and Building Exodus.
- **Structural Fire Performance:** Assessing how fire affects load-bearing elements and designing fire-resistant materials.
- **Fire Suppression Design:** Selecting appropriate sprinkler and gas suppression systems based on risk levels.

Integration of Fire Risk Assessment and Fire Engineering

Fire risk management is most effective when fire risk assessors and fire engineers collaborate.

- Fire risk assessments inform fire engineering design by identifying vulnerabilities.
- Fire engineers validate risk assessments using quantitative analysis and simulations.
- Joint efforts ensure compliance with regulations while optimizing safety and cost-effectiveness.

Case Studies of Fire Risk and Engineering Collaboration
- **Grenfell Tower Fire, UK:** Lessons learned in fire risk assessment failures and engineering solutions.

- **Notre-Dame Cathedral Fire, France:** Fire risk assessment for heritage structures.
- **High-Rise Evacuation Strategies in Dubai:** How fire engineering models influenced safer evacuation planning.

Regulatory Framework and Compliance

Fire risk management operates within a framework of national and international fire safety regulations:

- **UK:** Regulatory Reform (Fire Safety) Order 2005, Approved Document B, BS 9999, Fire Safety Act 2021, Building Safety Act 2022.
- **USA:** NFPA codes and standards (e.g., NFPA 101, NFPA 13, NFPA 92).
- **International:** ISO 23932 (Fire Safety Engineering), Eurocodes for structural fire design.

Global Fire Safety Harmonization

Harmonizing international fire safety standards ensures consistency in fire protection measures worldwide.

Digital Record-Keeping for Compliance

- **Blockchain-Based Fire Safety Logs:** Secure and immutable fire assessment records.
- **Cloud-Based Fire Safety Management Systems:** Remote monitoring and audit capabilities.

Fire risk management requires a multidisciplinary approach where fire risk assessors and fire engineers work together to mitigate hazards, enhance life safety, and ensure compliance with fire regulations. The integration of risk assessment methodologies and engineering solutions is essential for achieving fire-safe environments. With emerging technologies and regulatory advancements, fire risk management continues to evolve to meet modern fire safety challenges.

Methods for Assessing the Fire Risks

Types of Fire Risk Assessing and Their Methodologies

Fire risk assessments can be categorized into different types based on the level of detail and methodology applied:

1. Qualitative Fire Risk Assessment
2. Semi-Quantitative Fire Risk Assessment

3. Quantitative Fire Risk Assessment
4. PAS 79 Fire Risk Assessment
5. Fire Safety Engineering-Based Assessment

Each of these is discussed in detail below.

1. **Qualitative Fire Risk Assessment**
   - Focuses on subjective evaluation of fire hazards based on expert judgment.
   - Commonly used for small or low-risk premises.
   - Relies on checklists and visual inspections.
   - Methodology:
   - Identify hazards (e.g., ignition sources, fuel sources, oxygen supply).
   - Assess people at risk.
   - Evaluate existing control measures.
   - Recommend improvements and document findings.

2. **Semi-Quantitative Fire Risk Assessment**
   - Uses numerical scoring systems to rate fire risks.
   - More structured than qualitative assessments but still involves some subjectivity.
   - Suitable for medium-risk premises such as commercial offices and retail spaces.
   - Methodology:
   - Assign numerical values to hazard likelihood and consequences.
   - Calculate a risk rating (e.g., low, medium, high risk).
   - Prioritize risk mitigation actions.

3. **Quantitative Fire Risk Assessment**
   - Involves detailed statistical analysis and probabilistic modelling.
   - Commonly used for high-risk environments like industrial plants and high-rise buildings.
   - Uses fire growth models, evacuation simulations, and reliability analysis.
   - Methodology:
     1. Collect empirical data on fire incidents and system performance.
     2. Apply mathematical models (e.g., Monte Carlo simulations, fault tree analysis).
     3. Determine expected fire frequencies and consequences.
     4. Validate results using real-world case studies.

4. **PAS 79 Fire Risk Assessment**

PAS 79 and BS 9997 are key British standards that guide fire risk management and fire safety strategies in buildings.

PAS 79 (Publicly Available Specification 79) provides a structured methodology for conducting fire risk assessments. It is designed to help organizations comply with fire safety legislation by offering a standardized approach to identifying fire hazards, evaluating risk levels, and recommending appropriate control measures. PAS 79 has been widely used in the UK as a best-practice framework for ensuring fire safety compliance across various building types.

BS 9997:2019 is the British Standard for Fire Risk Management Systems (FRMS). It specifies requirements for implementing, maintaining, and improving a fire risk management system in alignment with ISO 31000 (Risk Management) and ISO 45001 (Health & Safety Management Systems). BS 9997 is particularly relevant for organizations that seek a systematic approach to managing fire risks beyond just compliance—integrating fire safety into broader business risk management strategies.

Both standards play a crucial role in promoting fire safety, with PAS 79 focusing on fire risk assessments and BS 9997 offering a comprehensive fire risk management framework for organizations looking to formalize their fire safety responsibilities.

5. **Fire Safety Engineering-Based Assessment**
   - Involves the application of engineering principles to fire risk analysis.
   - Suitable for complex buildings where prescriptive codes are not sufficient.
   - Uses CFD modelling, evacuation simulations, and performance-based design approaches.

Methodology

- Model fire dynamics using computational tools (e.g., FDS, SmartFire).
- Analyse human behaviour in fire scenarios.
- Design fire protection systems based on risk assessment outcomes.
- Validate the design through simulations and empirical data.

Regulatory Framework and Compliance

Fire risk management operates within a framework of national and international fire safety regulations:

- UK: Regulatory Reform (Fire Safety) Order 2005, Approved Document B, BS 9999

- USA: NFPA codes and standards (e.g., NFPA 101, NFPA 13, NFPA 92)
- International: ISO 23932 (Fire Safety Engineering), Eurocodes for structural fire design.

Fire risk management requires a multidisciplinary approach where fire risk assessors and fire engineers work together to mitigate hazards, enhance life safety, and ensure compliance with fire regulations. The integration of risk assessment methodologies and engineering solutions is essential for achieving fire-safe environments.

Chapter 8

Building Evacuation and Human Behaviour

EGRESS AND LIFE SAFETY DESIGN CONCEPT

Considering evacuation principles in building design is crucial and should be considered and tested at the design stage of the construction:

Safety

The primary goal of evacuation principles are to ensure the safety of building occupants during emergencies. A well-designed egress system can facilitate a quick and orderly evacuation, reducing the risk of injury or loss of life.

Regulatory Compliance

Building codes and regulations, such as the Building Safety Act 2022 in the UK, mandate the inclusion of specific egress and life safety features in building design. Non-compliance can result in legal penalties and increased liability.

Accessibility

Evacuation principles take into account the needs of all occupants, including those with disabilities. This ensures that everyone, regardless of their physical abilities, can safely evacuate the building in an emergency.

Efficiency

Effective evacuation design can help prevent bottlenecks and confusion during an emergency, enabling a more efficient evacuation process.

ASET vs RSET Comparison

A common fire engineering tool to analyse the conditions in a building in the event of a fire is the ASET (Available Safe Egress Time) vs. RSET (Required Safe Egress Time) comparison. The basic aim of this approach is to show that the calculated time before conditions become untenable (ASET) always exceeds the required time to safely evacuate the building (RSET). The time difference between the ASET and RSET is considered the safety margin, this must be large enough to allow for unseen hazards.

Evacuation Modelling

Evacuation modelling is a critical aspect of safety planning in building design. It involves the use of mathematical models and computer simulations to predict the behaviour of individuals and crowds during evacuation scenarios. These models take into account various factors such as the layout of the building, the number and location of exits, and the characteristics of the occupants (e.g., mobility, familiarity with the building). The goal is to identify potential bottlenecks, estimate evacuation times, and ultimately design more efficient and safer egress systems. Evacuation modelling can also be used to train occupants and emergency responders, and to evaluate the effectiveness of evacuation procedures and strategies. By simulating different scenarios, designers and safety officials can proactively address potential issues, thereby enhancing the overall safety of the building occupants during emergencies.

By considering these principles during the design phase, architects and engineers can create buildings that are not only functional and aesthetically pleasing, but also safe for all occupants.

Elements of egress and life safety design in the UK

Exit Access

This is the path that leads to an exit. It must be clear and unobstructed to facilitate easy movement towards the exits. The design should consider the layout of the building, furniture placement, and potential obstructions to ensure a clear path.

Exit Routes

These are the paths that occupants follow to evacuate the building. They should be well-marked and easily identifiable. The design should consider the number and location of exits, the capacity of each exit, and the distance to each exit.

Exit Discharge

This is the final part of the egress system that leads occupants to a safe location away from the building. The design should consider the location of the safe area, the capacity of the safe area, and the path to the safe area.

Emergency Lighting

This ensures visibility in low-light conditions, particularly during power outages. The design should consider the placement and intensity of emergency lighting, the duration of emergency lighting, and the reliability of the emergency lighting system.

Evacuation Strategies

These can include full or partial evacuation using stairs or lifts, egress for people with disabilities, protect-in-place strategies, and alternatives to evacuation. The design should consider the characteristics of the occupants, the nature of the building, and the potential emergencies that could occur.

Compliance with Regulations

Regularly reviewing and updating egress plans to accommodate changes in building layout or occupancy is crucial. Exit doors should be easily operable without special knowledge or tools, and clear signage and way finding should guide occupants to exits.

Building Safety Act 2022

This legislation provides the legal framework for building safety in the UK. It defines the responsibilities of building owners and managers, sets out the regulatory regime for building safety, and establishes the Building Safety Regulator.

By understanding these elements and implementing them effectively, we can enhance the safety of occupants and minimize risks during emergencies. It's important to note that the design of egress systems should be based on a thorough understanding of human behaviour during emergencies. This ensures that the egress design aligns with the likely actions of occupants during a fire or other emergencies.

Human Behaviour and Physiological Response to Fire

Understanding human behaviour during fire emergencies is crucial for effective evacuation planning and safety measures. Contrary to popular belief, emergencies often bring out our better nature, and human behaviour during such situations can be predicted and controlled through planning and understanding.

Panic and Behaviour

It is often wrongly assumed that panic during fires brings out the worst in people. However, research suggests that people are likely to apply rational, logical, and altruistic responses based on their understanding of the situation at the time of a fire.

Decision Making

Decisions made during a fire may not seem optimal in retrospect but were rational and the best ones considering all factors at the time.

Perception of Panic

The notion of panic during a fire is often influenced by the outcome of the fire. For instance, re-entering a burning building is labelled as "panic" if it results in fatality, but "heroic" if it results in lives saved.

Behaviour Patterns

According to Quarantelli's (1980) notion, there are five patterns of behaviour shown by people during fires and emergencies: warning, withdrawal, movement, shelter, and return.

Quarantelli's (1980) notion outlines five patterns of behaviour that people typically exhibit during fires and emergencies.

- **Warning**

This is the initial phase where individuals become aware of the emergency situation. They may receive this information from various sources such as fire alarms, smell of smoke, or word of mouth. The warning phase is crucial as it triggers the subsequent actions.

- **Withdrawal**

Upon recognizing the danger, individuals may attempt to remove themselves from the threatening situation. This could involve leaving the building or moving away from the source of danger.

- **Movement**

This phase involves the actual process of evacuation. It's important to note that people are more likely to use familiar routes during evacuation. The movement can be influenced by various factors including the design of the building, the intensity of the fire, and the density of smoke.

- **Shelter**

In some cases, individuals may seek shelter instead of evacuating, especially if they perceive the outside environment to be more dangerous. This could involve staying in a safe room within the building until help arrives.

- **Return**

After the danger has passed, individuals may return to the affected area to assess the damage or retrieve their belongings. However, this phase can be risky if the structure has been weakened by the fire.

It's important to note that these behaviours are not strictly sequential and can overlap. For instance, an individual might be in the movement phase but may need to seek shelter if their evacuation route is blocked. Understanding these patterns can help in designing effective fire safety measures and evacuation plans.

Evacuation Strategies

In the UK, evacuation strategies for buildings, particularly high-rise residential buildings, are a critical aspect of fire safety. The Home Office and the University of Central Lancashire (UCLan) have conducted extensive research and live testing to develop effective evacuation strategies.

The research focused on five key strategies:

Strategy 1

A full simultaneous evacuation using an Evacuation Alert System (single staircase).

This strategy was one of the five tested during the live operational tests conducted between May 3 and 6, 2022. The results indicated that using an Evacuation Alert System resulted in faster evacuation than a door-knocking system. In fact, evacuees using the door-knocking system left the building more than eight minutes longer than those using the Evacuation Alert System. These findings suggest that an Evacuation Alert System can significantly improve the efficiency of evacuation in high-rise residential buildings.

Strategy 2

A full evacuation with door-knocking alerts from bottom to top of the building, without an Evacuation Alert System (single staircase).

This strategy involved manually alerting residents to evacuate by knocking on doors, starting from the bottom of the building and moving upwards. The findings indicated that this door-knocking system was less efficient than using an Evacuation Alert System. These results highlight the importance of efficient alert systems in ensuring timely evacuation in high-rise residential buildings.

Strategy 3A

A full evacuation using an Evacuation Alert System (single staircase): phased bottom-up from above the fire.

This strategy involved using an Evacuation Alert System (EAS) to alert residents to evacuate in a phased manner, starting from the floor above the fire and moving upwards. it was noted that the use of an EAS generally resulted in faster evacuation times compared to manual door-knocking alerts. The specific efficiency of this phased bottom-up strategy would depend on various factors such as the location of the fire, the building layout, and the responsiveness of the residents.

Strategy 3B

A full evacuation using an Evacuation Alert System (single staircase): phased top-down from above the fire.

This strategy also involved using an Evacuation Alert System (EAS) to alert residents to evacuate in a phased manner, starting from the top floor and moving downwards. The efficiency of this phased top-down strategy would also depend on various factors such as the location of the fire, the building layout, and the responsiveness of the residents.

Strategy 4

A full simultaneous evacuation using an Evacuation Alert system with two staircases.

This strategy involved using an Evacuation Alert System (EAS) to alert all residents to evacuate simultaneously, utilizing two staircases. The findings indicate that two staircases facilitate faster evacuation compared to a single staircase, with the EAS proving more efficient than manual door-knocking alerts. The efficiency of this strategy would depend on various factors such as the building layout, the location of the staircases, and the responsiveness of the residents

These strategies were tested in live operational scenarios involving fire and rescue service staff and volunteers acting as residents. The tests aimed to answer research questions related to the efficiency of different evacuation strategies, the impact of congestion in stairwells, the effect of evacuees with impairments, and other factors that could affect live evacuations.

In addition to these strategies, the Home Office has set out nine national guidelines to support operational guidance and practices during a full or partial evacuation from high-rise residential buildings. These guidelines were developed in response to the Grenfell Tower Inquiry and are based on recent research, including live evacuation exercises and studies on human behaviour and public confidence.

Building Height (Timing of Evacuation)

The total evacuation time in the stairwell should not be derived from travel distance alone. Doubling the building height does not necessarily double the total evacuation time.

Staircases (Timing of Evacuation)

There is a benefit of having an unobstructed staircase for the sole use of evacuation in a building during a fire incident.

Evacuation Alarm Systems versus Door-Knocking

The efficiency of evacuation alert systems compared to manual door-knocking alerts.

Emergency Evacuation Methods for Residents

Different strategies for resident evacuation.

Evacuation Movement and Vulnerable Residents

The impact of evacuees with impairments on overall evacuation times and the potential for congestion in stairwells.

Undertaking Evacuations of Vulnerable Residents

The procedures for evacuating persons who are unable to use the stairs in an emergency, or who may require assistance.

Information on Vulnerable Residents

The need for information on residents who may require assistance during an evacuation.

Evacuees' Behaviour

Understanding resident behaviour during fire incidents.

Information Sharing Amongst Residents

The importance of clear instructions and information sharing among residents during an incident.

Evacuation Analysis

Fire engineered analysis of evacuation is a method that simulates a real-life fire event to predict evacuation times for areas and buildings. It is based on how people will react under the pressure of a fire, when emotion, possible crowding, and the response of the materials and systems within the building come into play. This section builds on the earlier section which describes the different evacuation strategies.

Key aspects of a fire engineered analysis of evacuation include:

Evacuation Modelling

This involves the use of computational models to simulate the evacuation process during a fire scenario. These models can help assess the safety levels of specific designs and compare the safety levels for alternative designs.

Human Behaviour

The models consider human behaviour under the pressure of a fire, including factors such as emotion and possible crowding.

Building Response

The response of the materials and systems within the building during a fire is also considered in the analysis.

Innovative Design

Fire engineered analysis of evacuation provides opportunities for innovative design, allowing for the creation of buildings that are not only aesthetically pleasing but also safe.

Management of Fire Safety

The analysis provides information on the management of fire safety for a building.

In the context of BS 7974:2019, a fire engineered analysis of evacuation would be part of the overall fire safety engineering approach. This approach applies scientific and engineering principles to protect people, property, and the environment from fire. It is applicable to the design of new buildings and the appraisal of existing buildings.

ASET vs. RSET

The Available Safe Egress Time (ASET) and Required Safe Egress Time (RSET) are fundamental concepts in fire safety engineering, particularly for assessing the adequacy of building egress systems. This chapter provides a detailed examination of these parameters, including the use of B-Risk for ASET calculations and comprehensive egress calculations.

ASET (Available Safe Egress Time)

ASET refers to the duration from the initiation of a fire to the point at which conditions become untenable for occupants. It is determined through fire modelling, including smoke movement, heat release rates, and toxicity levels.

Use of B-Risk for ASET Calculations

B-Risk is a zone model developed (by BRANZ Building Research Association of New Zealand) to predict fire and smoke behaviour in compartment fires. It considers:

- Fire growth rate
- Smoke production and layer height
- Visibility limits due to smoke
- Toxic gas concentrations
- Tenability criteria (temperature, visibility, toxicity thresholds)

Calculation Example using B-Risk

- **Fire growth rate assumption:** Medium growth rate with HRR of 0.047 kW/s²
- **Smoke layer descent:** Using B-Risk model outputs
- **Visibility criteria:** Minimum of 10m for escape route
- **Toxicity:** Fractional effective dose (FED) for CO ≤ 0.3

ASET Determination

From B-Risk outputs, ASET is typically determined when visibility drops below 10m, or toxic levels exceed limits.

RSET (Required Safe Egress Time)

RSET is the time required for all occupants to evacuate safely. It comprises multiple stages:

$$t_{evac} = t_{pre} + t_r + t_q$$

Where

t_{evac} is the RSET, the required evacuation time given in minutes

t_{pre} is the pre-movement time

t_{tr} is the travel time given in minutes

t_q is the queuing time

Evacuation Time

Pre-Movement Time

Pre-movement time includes the reaction delay from fire detection to the start of movement. It varies based on:

- Awareness (awake/asleep)
- Fire detection systems (automatic vs. manual)
- Occupant characteristics

Pre-Movement time is given as

$$t_{pre} = +t_a + t_{r+d}$$

Where

t_{pre} is the time from when the fire starts until the occupants begin to evacuate

t_a is the time from when the fire starts to when it is detected, and occupants are alarmed

t_{r+d} is the time from the alarm raised to occupants beginning evacuation

Typical values:

Occupant Type	Pre-movement Time (seconds)
Awake, alert	30-60
Asleep	120-300
Disabled	180+

Travel Time

Travel time depends on distance, walking speed, and obstacles.

Travel time is given as

$$t = \frac{L_{tr}}{S}$$

Where

t_{tr} is the travel time given in minutes

L_{tr} is the distance, or length of travel

S is the speed of travel

Walking Speeds by Age:

Age Group	Speed (m/s)
Children	0.9-1.1
Adults	1.2-1.4
Elderly	0.8-1.0

Speed Reduction Due to Density:

Density (persons/m²)*	Speed Reduction (%)
0.5	0%
1.0	10%

Density (persons/m²)*	Speed Reduction (%)
2.0	40%
3.0	60%

*Occupant density was used from BS9999:2017

The geometry of the room can also have a bearing on the speed of travel. The speed can then be determined using the formula:

$$S = \min\{\tfrac{72.828}{k_t(1-0.266D_o)}\}[m/\min]$$

Where

S is the speed of travel

D_o is the occupant density

k_t depends on the geometry of the particular location being investigated. This can be given as:

Level corridors or doorways $k_t = 84.0$, and

$$k_t = 51.8\left(\frac{G}{R}\right)^{0.5}$$

Stairs

Queuing Time

Queuing occurs at bottlenecks such as doorways and stairwells.

Queue Time Calculation

$$t_q = \frac{N}{F_a}$$

Where:

t_q = queuing time

N = number of people

F_a = flow rate (people/s)

Acceptance Criteria

Evacuation Safety Level

The parameters to consider when making an evacuation analysis are.

RSET: Required Safe Egress Time. The time required for the occupants to safely evacuate

RSET = t_{evac}

ASET = Available Safe Egress Time. The time until untenable conditions occur based on various acceptance criteria (see following formulas)

$$ESL = \frac{ASET}{RSET}$$

Where

ESL is the Evacuation Safety Level

$ASET$ is the Available Safe Egress Time

$RSET$ is the Required Safe Egress Time

The acceptance criteria for occupant safety is based on the evacuation taking place before the conditions become untenable.

Generally, the evacuation should be completed before the ASET conditions in Table below are met.

Parameter	ASET	RSET
Visibility Limit	10m	N/A
Temperature	>80°C	N/A
Toxicity Threshold	FIC ≤ 1	FID ≤ 1
Time (seconds)	180	118

This criterion must be met in a free height, Z (m) above the floor. This height is determined based on the height, H (m) of the room.

Fractional Irritant Concentration (FIC)

FIC assesses exposure to irritants such as CO and HCN which are found in smoke. $C_{irritant}$ is the concentration which is determined to impair escape.

Gas	Symbol	$C_{irritant}$
Sulphur dioxide	SO_2	24 ppm
Nitrogen dioxide	NO_2	70 ppm
Hydrogen chloride	HCl	200 ppm
Hydrogen bromide	HB_r	200 ppm

Gas	Symbol	$C_{irritant}$
Hydrogen fluoride	HF	200 ppm
Hydrogen cyanide	HCN	50 ppm
Acrolein	CH_2CHO	4 ppm
Formaldehyde	HCHO	6 ppm
Carbon monoxide	CO	1500 ppm

The fractional irritant concentration (FIC) is the actual concentration over the irritant concentration, where FIC = 1 is expected to impair escape for 50% of the population, and FIC = 3-5 is expected to cause incapacitation.

The severity of various gases can be made using the following formula

$$FIC_{mix} = \sum_{j=1}^{N} \frac{C_j}{C_{i,j}}$$

Where

FIC_{mix} is the fractional incapacitating dose of the irritant
C_j is the actual concentration of the species j
$C_{i,j}$ is the irritant concentration of the species j from the table above
N is the total number of species
The conditions are unacceptable if FIC ≤ 1

Fractional Incapacitating Dose (FID)

FID evaluates exposure to toxic gases.

$$FID = (F_{1,CO} + F_{1,CN} + F_x)V_{CO2} + F_{1,O2}$$

Where

FID is the Fractional Incapacitating Dose
F_x is the fractional dose of irritants causing asphyxia
$F_{1,CO}$ is the fractional dose of CO causing asphyxia
$F_{1,CN}$ is the fractional dose of cyanide causing asphyxia
V_{CO2} is the factor taking into account hyperventilation due to elevated CO_2
$F_{1,O2}$ is the oxygen depletion factor

Formula	Input
$F_{1,CO} = \dfrac{8.2925 \cdot 10^{-4} C_{CO}^{1.036}}{30}$	PPM
$F_{1,CN} = \dfrac{\left(\exp(C_{HCN} + C_{othernitriles} - C_{NO2})\right)^{1/43}}{220}$	PPM
$V_{CO2} = \dfrac{\exp(C_{CO2})}{5}$	%
$F_{1,O2} = \dfrac{1}{\exp(8.13 - 0.54(20.9 - C_{O2}))}$	%

As displayed C_i is the concentration of species i in the given input unit.

Example RSET Calculation

- **Pre-movement time:** 60s
- **Travel time:** 30m distance at 1.2 m/s = 25s
- **Queue time:** 50 people, flow rate 1.5 p/s = 33s

  Total RSET = 60 + 25 + 33 = 118s

Comparison of ASET vs. RSET

Parameter	ASET	RSET
Visibility Limit	10m	N/A
Toxicity Threshold	FIC ≤ 1	FID ≤ 1
Time (seconds)	180	118

If ASET > RSET, the design is considered safe.

Understanding ASET and RSET is essential in fire engineering. By using B-Risk (in this example) for ASET predictions and considering detailed egress calculations, designers can ensure compliance with safety standards and mitigate fire risks effectively.

Chapter 9

Fire Protection Systems

Active Fire Protection (AFP)

Active fire protection refers to systems or measures that require some level of human intervention, mechanical action, or automated response to detect, suppress, or control a fire. These systems are designed to actively engage when a fire is detected, either through automated triggers or manual activation. AFP is essential for controlling fires at their onset and preventing them from spreading.

Components of Active Fire Protection

Fire Detection Systems

These systems identify the presence of a fire early, allowing for a prompt response. The most common types of fire detection systems include:

- **Smoke detectors**: Sense the presence of smoke.
- **Heat detectors**: Respond to temperature increases.
- **Flame detectors**: Detect the presence of flames in sensitive environments.

Fire Suppression Systems

These systems work to actively extinguish or suppress a fire once it has been detected. They include:

- **Sprinkler systems**: Water-based systems that automatically activate when the heat from a fire reaches a certain level.
- **Foam suppression systems**: Used in high-risk environments where flammable liquids are present, these systems disperse foam to smother the fire.
- **Gas suppression systems**: Systems that release gas (like carbon dioxide or clean agents) to suppress fires in sensitive environments (e.g., server rooms).
- **Water mist systems**: Use fine water droplets to cool the flames and surrounding air, reducing heat and suffocating the fire.

Manual Firefighting Equipment

Fire extinguishers: Portable devices that can be manually operated to extinguish small fires. Different types include:

- **Water extinguishers:** Fires involving solid combustibles like wood, paper, textiles (Class A fires).
- **Foam extinguishers:** Fires involving solids (Class A fires) and flammable liquids like petrol and diesel (Class B fires)
- **CO_2 extinguishers:** Electrical fires and flammable liquids (Class B fires).
- **Wet Chemical Extinguishers:** Fires involving cooking oils and fats (Class F), e.g., in deep-fat fryers for electrical or cooking fires (Class F).
- **Dry Powder Extinguishers:** Solid Combustibles (Class A) Flammable Liquids (Class B) Flammable Gases (Class C)

Fire Alarm Systems

Once a fire is detected, alarms are triggered to alert building occupants and emergency responders. Modern fire alarm systems are often connected to monitoring centres, ensuring a rapid emergency response.

Firefighter Access

Standpipes and fire hoses provide easy access for firefighters to use water within a building, ensuring they can quickly combat a fire upon arrival.

Emergency Lighting and Evacuation Systems

Active fire protection systems also include emergency lighting and audible evacuation alarms to guide building occupants to safety during a fire.

Advantages of Active Fire Protection

Immediate response

Systems such as sprinklers and alarms activate almost instantly once a fire is detected, providing quick suppression and alerting people to evacuate.

Flexibility

Active fire protection systems can be tailored to specific environments (e.g., foam for industrial areas, gas suppression in data centres).

Life-saving potential

Early detection and suppression systems are critical for saving lives and minimizing injury during fire incidents.

Disadvantages of Active Fire Protection

Maintenance-dependent

Active systems require regular inspection, maintenance, and testing to ensure they function correctly during an emergency.

Potential for malfunction

In the event of poor maintenance, components may fail to activate or may activate unnecessarily (e.g., false alarms or sprinkler discharge).

Cost of installation

Some systems, such as gas suppression or foam systems, can be costly to install and maintain.

Passive Fire Protection (PFP)

Passive fire protection

Refers to the built-in elements of a structure designed to contain or slow the spread of fire. Unlike active systems, passive fire protection is always "on" and doesn't require activation to function. These systems help to compartmentalize the fire, protect escape routes, and maintain the structural integrity of the building for a longer period, giving occupants more time to evacuate and firefighters more time to respond.

Key Components of Passive Fire Protection

Fire-Resistant Walls and Doors

Fire-resistant walls and partitions: These are designed to contain the spread of fire by creating barriers between different sections of a building.

Fire-rated doors: These doors help prevent the fire and smoke from spreading through corridors and exit routes, containing the fire to a specific area.

Fire Compartmentation

The practice of dividing a building into compartments (rooms or zones) that are separated by fire-resistant barriers. This slows the spread of fire, smoke, and heat, providing a safe pathway for evacuation.

Fireproofing Materials

Fireproof coatings and claddings:

These materials are applied to structural components, such as steel beams and columns, to insulate them from the heat of a fire. This helps prevent the structure from collapsing under the intense heat.

Intumescent coatings

Special paints or coatings that expand when exposed to heat, creating an insulating layer that protects structural elements from fire damage.

Fire-Resistant Glazing

Fire-rated glass is used in windows and doors to prevent the spread of fire between compartments while maintaining visibility. These types of glass can withstand high temperatures and remain intact during a fire.

Smoke Barriers and Dampers

Smoke barriers: Designed to limit the spread of smoke through a building by dividing it into compartments.

Fire dampers: Installed within ventilation systems to close automatically when a fire is detected, preventing smoke and flames from spreading through ducts.

Fire stopping Materials

Used to seal gaps around penetrations in fire-rated walls, such as piping, wiring, and ductwork, preventing the spread of fire through small openings.

Fire-Rated Ceilings and Floors

Fire-resistant ceilings and floors: These components help slow the vertical spread of fire, protecting upper and lower floors from collapse or damage.

Advantages of Passive Fire Protection

No manual intervention required

Passive systems do not rely on activation or human involvement—they are built into the structure and work continuously.

Long-term effectiveness

Once installed, passive systems provide consistent protection with minimal maintenance (compared to active systems).

Containment

PFP helps contain fires, reducing the amount of damage and preventing the fire from spreading throughout the building.

Support for active systems

Passive measures give more time for active fire protection systems, firefighters, and occupants to react.

Disadvantages of Passive Fire Protection

Limited fire control

While passive systems contain or slow the fire, they don't actively extinguish it, meaning the fire will continue to burn until suppressed by other means.

Potential design limitations

If poorly planned, PFP might not be as effective, especially in buildings with complex designs or without proper compartmentalization.

Renovation challenges

Installing or upgrading passive systems in existing buildings can be complex and expensive, especially when retrofitting fire-rated barriers or structural elements.

Key Differences Between Active and Passive Fire Protection

		Active Fire Protection Systems	Passive Fire Protection Systems
1	**Functionality**	Actively engage to control or suppress a fire (e.g., sprinklers, fire alarms)	Contain the fire and limit its spread without any activation or moving parts (e.g., fire-rated walls, doors)
2	**Response to Fire**	Require detection and activation to function, either automatically or manually	Are integrated into the building's structure and work without needing any activation
3	**Maintenance**	Require regular testing, maintenance, and, sometimes, user interaction to ensure they work properly in emergencies	Typically require less frequent maintenance but must be installed correctly and up to code to be effective
4	**Objective**	Are designed to detect, control, and suppress fires	Are designed to contain, compartmentalize, and slow the spread of fires.

Both active and passive fire protection systems play crucial roles in overall fire safety strategy. Active fire protection provides the first line of defence by detecting and responding to a fire, while passive fire protection ensures that a fire's spread is contained, giving occupants time to evacuate and emergency responders time to act.

A well-rounded fire protection plan includes both active and passive systems working together to mitigate fire risks, minimize damage, and ensure occupant safety.

Sprinkler Systems

Sprinkler Systems are a critical component of active fire protection designed to detect, control, and suppress fires automatically. They are highly effective at reducing fire damage and preventing the spread of flames, providing valuable time for building occupants to evacuate and for firefighters to respond. Below is a detailed account of sprinkler systems, including their components, types, design considerations, and operational mechanisms.

Overview of Sprinkler Systems

Sprinkler systems are composed of networks of water pipes, valves, and sprinkler heads that discharge water when a fire is detected. They are among the most widely used fire suppression systems in commercial, industrial, and residential buildings due to their reliability and effectiveness.

Key Components of a Sprinkler System

Sprinkler Heads

- **Sprinkler heads** are the outlets from which water is discharged when a fire is detected. Each head contains a heat-sensitive element that activates the sprinkler when it reaches a certain temperature. Types of sprinkler heads include:
    - **Pendant heads**: Hang from the ceiling and spray water downward.
    - **Upright heads**: Designed for areas where obstructions might block downward spray, these heads release water upward, deflecting off a plate to spread water.
    - **Sidewall heads**: Mounted on walls to spray water in a semicircular pattern, often used in small spaces or where ceiling installations are difficult.

Heat-Sensitive Activation Mechanisms

- Sprinkler heads are equipped with either:
    - **Glass bulbs**: Contain a liquid that expands when heated, causing the bulb to burst at a specific temperature (usually between 135°F and 165°F, but higher ratings are available for special conditions).
    - **Fusible links**: Two metal pieces are soldered together. When the surrounding temperature rises, the solder melts, allowing the sprinkler to activate.

- **Water Supply**

The water supply system provides the pressure and volume needed to operate the sprinklers. It typically consists of:

- **Water storage tanks**: Used in buildings where a reliable water supply is not guaranteed.
- **Water mains**: Pipes connected to a municipal or private water supply system.
- **Pumps**: Boost water pressure when needed, ensuring adequate flow to the sprinklers.

Piping Network

- **Pipes** deliver water from the main supply to the sprinkler heads. The pipes are usually designed to withstand high pressure and are configured to cover the entire protected area. Materials include:
  - **Steel** (most common),
  - **copper**, or
  - high-performance **plastic** pipes in certain applications.
- **Control Valves**

  Control valves regulate the flow of water to the system and are typically located near the water supply connection. Key valves include:
  - **Main control valve**: Must be in the open position to allow water to flow. It is equipped with a tamper switch that triggers an alarm if the valve is closed.
  - **Alarm valve**: Activates the fire alarm system when water starts to flow through the pipes.
  - **Zone control valves**: Allow separate areas or zones of the building to be isolated or shut off without affecting the entire system.

Fire Pumps

- In buildings where the water supply pressure is insufficient, **fire pumps** are installed to ensure that water can reach the sprinkler heads at adequate pressure. Types of pumps include electric-driven and diesel-driven pumps.
- **Alarm Systems**
  - When a sprinkler system is activated, it triggers a connected fire alarm system, notifying building occupants and emergency services. Alarms can also activate water flow indicators to track which part of the system has been triggered.

Types of Sprinkler Systems

There are several types of sprinkler systems, each suited to different building environments and fire risks. The main types include:

1. **Wet Pipe Sprinkler System**

   This is the most common and straightforward type of sprinkler system. In wet pipe systems, the pipes are always filled with water under pressure, allowing for an immediate discharge of water when a sprinkler head activates.

2. **Dry Pipe Sprinkler System**

   In dry pipe systems, the pipes are filled with pressurized air or nitrogen instead of water. When a sprinkler head activates, the air is released, allowing water to flow through the pipes to the sprinkler heads. This delay in water discharge is minimal but necessary in environments where pipes could freeze.

3. **Pre-Action Sprinkler System**

   Pre-action systems require two triggers to activate. First, a separate detection system (such as smoke or heat detectors) must sense a fire. This opens the pre-action valve, allowing water to fill the pipes. The sprinklers will only discharge water if a sprinkler head then activates due to heat from the fire.

4. **Deluge Sprinkler System**

   In deluge systems, all sprinkler heads are open, but the pipes are dry. When a fire is detected (via a fire detection system), water is released through all the sprinkler heads simultaneously. These systems are used in high-hazard environments where rapid and complete fire suppression is critical.

5. **Foam Water Sprinkler System**

   Foam water systems mix water with foam concentrate, creating a foam that suppresses fire by cooling and smothering it. Foam is especially effective for fires involving flammable liquids and other high-risk materials.

6. **Water Mist Sprinkler System**

   Water mist systems use very fine droplets of water, which absorb heat more efficiently than traditional sprinklers. The mist cools the flames and reduces the oxygen around the fire, helping to extinguish it.

The advantages, dis-advantages and applications of each of these Sprinkler Systems are given in the table below:

	System	Advantages	Dis-advantages	Applications
1	Wet Pipe Sprinkler system	• Quick response since the water is already in the pipes. • Simple design and low maintenance.	• Not suitable for environments prone to freezing temperatures (e.g., outdoor areas or unheated buildings).	Office buildings, hotels, residential properties, and shopping centres.
2	Dry Pipe Sprinkler System	• Suitable for unheated buildings or cold environments. • Protects pipes from freezing damage.	• Slight delay in water discharge compared to wet pipe systems. • More complex and costly to install and maintain.	Warehouses, parking garages, or industrial facilities located in cold climates.
3	Pre-Action Sprinkler System	Reduces the risk of accidental water discharge (which is especially important in sensitive environments like data centres or museums). Allows aa second verification (via a detector) before water is released.	More expensive and complex to install and maintain. Delays water discharge, though minimal when properly configured.	Data centres, libraries, museums, and areas with sensitive or high-value contents.
4	Deluge Sprinkler System	Provides immediate, large-scale water discharge to suppress fires. Particularly effective in high-risk areas with rapid fire spread potential.	Risk of significant water damage. Requires precise control and detection systems.	Power plants, chemical processing facilities, airplane hangars, and refineries.

5	Foam Water Sprinkler System	Superior suppression for fires involving flammable liquids. Effective for preventing fire re-ignition.	Requires special foam agents and maintenance. May leave residue that requires clean-up after fire suppression.	Aircraft hangars, chemical storage facilities, fuel stations, and hazardous material storage areas.
6	Water Mist Sprinkler System	Uses less water, minimizing water damage. Effective for areas where water damage must be minimized, such as historical buildings.	More complex design and installation. Higher installation costs than conventional systems. Should risk assessed before a decision on its use* *See Section on smoke species in chapter "Fire Engineering Science"	Historic buildings, museums, data centres, and areas sensitive to water damage.

Design Considerations for Sprinkler Systems
Fire Hazard Classification

Buildings are categorized based on the level of fire hazard they present. These classifications guide the design of sprinkler systems, determining water pressure, flow rates, and the type of sprinkler heads used. The main categories include:

- **Light hazard**: Low fire load environments like offices or schools.
- **Ordinary hazard**: Moderate fire load environments like retail stores, garages, or manufacturing plants.
- **High hazard**: High fire load environments like chemical plants, warehouses, or fuel storage facilities.

Water Supply and Pressure

The availability and reliability of the water supply are crucial. Designers must ensure that the system can provide enough water at sufficient pressure during an emergency. If the building's water supply is insufficient, storage tanks or fire pumps may be necessary.

Building Occupancy and Layout

The number of occupants, building layout, and evacuation routes are critical factors. Sprinkler coverage must be designed to ensure that fires are controlled long enough for safe evacuation.

Local Codes and Regulations

Sprinkler systems must comply with local fire codes and building regulations, which set requirements for system design, installation, and maintenance.

Maintenance and Testing

To ensure that sprinkler systems work effectively during a fire, regular maintenance and testing are critical. This includes:

- **Weekly, monthly, and annual inspections** to check valves, pumps, and sprinkler heads.
- **Flow testing** to ensure adequate water pressure and discharge rates.
- **Tamper alarms** testing to ensure that control valves are not inadvertently closed.
- **Sprinkler head replacement**: Older heads

Fire Alarm and Detection Systems

In the UK, fire alarm and detection systems must comply with various codes and regulations to ensure they meet safety standards, provide adequate protection, and help in the effective evacuation of building occupants. These codes and regulations set out the requirements for the design, installation, maintenance, and testing of fire alarm systems, and they are aligned with broader fire safety laws.

Regulatory Framework For Fire Safety In The UK

The Regulatory Reform (Fire Safety) Order 2005 (RRO)

The Regulatory Reform (Fire Safety) Order 2005 (RRO) is the primary piece of legislation governing fire safety in non-domestic premises in England and Wales (it also applies to Scotland and Northern Ireland, with some variations). The RRO places a duty on employers and building owners to ensure the safety of all occupants from fire risks.

Key points relevant to fire alarms:

- A responsible person (employer, owner, or manager) must ensure a fire risk assessment is conducted to identify hazards, risks, and the necessary preventive and protective measures.
- The fire risk assessment must take into account the provision of fire alarms and detection systems appropriate to the type of premises and level of risk.
- Fire detection and alarm systems must be maintained in efficient working order and regularly tested.
- The system should be capable of alerting occupants to the presence of fire early, giving sufficient time for evacuation.
- The fire alarm system must comply with UK standards for the type of occupancy and risk level.

UK Standards for Fire Alarm Systems

The primary British Standard (BS) codes and regulations that apply to fire alarm systems are:

1 BS 5839-1: 2017 - Fire Detection and Fire Alarm Systems for Buildings: Part 1

This is the UK's main standard for the design, installation, commissioning, and maintenance of fire detection and fire alarm systems in non-domestic premises.

Key points include:

- **System design**

The fire alarm system must be designed to detect a fire and alert building occupants in a timely manner. The design will depend on the type of building, its occupancy, and associated fire risks.

Type of system

BS 5839-1 outlines the criteria for various types of systems, such as:

Manual systems: Systems where alarms are activated manually by a call point.

Automatic detection systems: Systems with detectors that automatically activate in the presence of smoke, heat, or gas.

Voice alarm systems: Systems that use voice messages to assist with evacuation instructions.

Alarm notification: Alarm devices, such as bells, horns, and strobe lights, must be installed to alert occupants, particularly in noisy environments or areas with hearing-impaired individuals.

Zoning: The system must be able to identify the location of the fire, either by zone or specific detector.

Maintenance: Regular testing and inspection of the fire alarm system to ensure it remains fully operational.

2. BS 5839-6: 2019 - Fire Detection and Fire Alarm Systems in Domestic Premises

This standard applies to fire alarm and detection systems installed in domestic premises (such as homes and flats).

Key points include:

- **Type of detectors**: For residential premises, smoke detectors are often recommended, while heat detectors may be used in areas like kitchens.
- **System design**: Unlike commercial systems, BS 5839-6 provides guidance on simpler, often battery-operated, systems that may only include smoke alarms in the most critical areas (e.g., hallways, landings, or escape routes).
- **Interlinked alarms**: Smoke alarms should be interlinked so that if one detects smoke, all alarms in the house will sound, providing a more effective warning for residents.

3. BS 5839-9: 2011 - Fire Detection and Fire Alarm Systems: Part 9

This standard focuses on the use of fire detection and alarm systems in healthcare premises, including hospitals and care homes.

Key points include:

- **Special considerations for healthcare environments**: Systems should be tailored to meet the needs of vulnerable individuals, such as people with mobility issues or cognitive impairments, with additional features like visual alarms and voice evacuation systems.

4. BS 5266-1: 2016 - Emergency Lighting

Though primarily dealing with emergency lighting, this standard is relevant for fire alarm systems because it integrates with the fire alarm system to ensure that evacuation routes are lit and safe in the event of a fire.

Considerations for Fire Alarm System Design

The design of a fire alarm system must consider several factors to comply with the Regulatory Reform (Fire Safety) Order 2005 and the relevant British Standards.

Risk Assessment

The fire alarm system must be designed based on the outcome of a fire risk assessment. This assessment identifies the building's fire hazards, the number of occupants, the layout of the building, and the fire protection measures needed. Different systems may be required for different types of buildings, from offices and factories to schools and hospitals.

Type of Premises and Occupancy

- **High-risk premises** (e.g., industrial sites, chemical plants) require more advanced fire alarm systems that are tailored to specific hazards.
- **Low-risk premises** (e.g., offices, shops) may require simpler systems.
- **Large, complex buildings** may need addressable systems where the location of each detector is easily identifiable, while smaller buildings may only require conventional systems.

Alarm Sounding and Visual Indicators

- Fire alarms must be loud enough to be heard throughout the premises, especially in areas with high ambient noise levels.
- **Visual signals** (e.g., flashing lights) should be installed where the hearing-impaired or people with hearing difficulties might be present.

Maintenance and Testing

Routine maintenance and testing must be conducted to ensure the fire alarm system is fully operational. This includes:

Weekly checks (e.g., ensuring that control panels are operational).

Annual inspections (e.g., testing detectors, alarm devices, and the power supply).

5. Other Relevant Fire Safety Regulations in the UK

In addition to the Regulatory Reform (Fire Safety) Order 2005 and the British Standards, there are other regulations that may influence the design and operation of fire alarm systems:

Health and Safety at Work Act 1974 (HSWA)

This is the overarching piece of legislation for health and safety in the UK. It places a duty on employers to ensure the safety of all employees and others who may be affected by their work activities, including providing adequate fire safety measures.

The Building Regulations 2010 (England and Wales)

The Building Regulations cover a broad range of safety issues, including fire safety. They require that all new buildings (or substantial renovations) provide adequate means of escape and fire detection.

Part B of the Building Regulations deals specifically with fire safety and outlines requirements for fire alarms, emergency lighting, escape routes, and fire-resisting materials.

Inspection, Testing, and Certification

Fire alarm systems must be regularly inspected and tested to ensure their continued reliability and functionality. The following requirements apply:

- Inspection and testing should be conducted by qualified professionals, often certified through the National Security Inspectorate (NSI) or the British Standards Institute (BSI).
- **Testing frequency**: At a minimum, systems should be tested annually, but specific maintenance tasks (such as checking batteries or ensuring detectors are clear of obstructions) may require more frequent testing.

Record keeping: Detailed records of all maintenance and testing activities must be maintained to demonstrate compliance with regulatory requirements.

Fire Alarm Systems

In the UK, fire alarm systems are categorized into different grades and categories according to the BS 5839-1:2017 standard. These categories are used to define the type of fire alarm system that should be installed based on the building's purpose, layout, and the level of fire risk. The classification involves both the category of system (L1, L2, L3, etc.) and the grade of system (A, B, C, etc.), which refers to the system's level of sophistication and coverage.

Categories of Fire Alarm Systems (BS 5839-1)

Category L Systems - Life Protection

These systems are designed to protect life by giving early warnings to people in a building, allowing for safe evacuation. The purpose, description and features of the various categories of Life Protection Systems, including a Manual System, are outlined in the table below.

	System Category		Purpose	Description
1	L1 – Full Coverage System	Provides complete coverage of the building, including all areas where people are likely to be present, such as offices, corridors, staircases, and rooms	This system is the most comprehensive, with fire detectors installed in all rooms, corridors, escape routes, and other areas where the risk to life is high. It is usually required for buildings with high fire risks or where large numbers of people gather, such as hotels, hospitals, and multi-story offices	Detectors in all rooms, staircases, corridors, and escape routes. Suitable for buildings with large numbers of people or high-risk environments. Provides the earliest possible warning of fire throughout the building
2	L2 – Cloned Coverage System	Provides early detection of fire in areas where people may be at risk, but not necessarily throughout the entire building.	This system includes detectors in critical areas such as escape routes, areas where fire risk is higher, and spaces like kitchens, boiler rooms, and electrical rooms. The system will not typically cover individual rooms unless they are deemed to be higher risk.	Detectors are installed in escape routes and other high-risk areas. May be suitable for schools, office buildings, and smaller or medium-sized businesses. Reduces the cost compared to L1 systems by excluding some lower-risk areas.

3	**L3 – Escape Route System**	Focuses on ensuring that **escape routes** are protected and provide warnings to people as they evacuate the building.	This system installs detectors in the escape routes (e.g., corridors, stairwells) and may exclude other areas of the building. The aim is to ensure that an evacuation can take place safely, even if fire or smoke is detected in the building.	Protection in escape routes and fire exit pathways. Appropriate for buildings with **lower occupancy** and less risk where the primary concern is ensuring people can evacuate safely. May be used in residential buildings or small offices.
4	**L4 – Property Protection Systematic**	Designed to provide **early fire detection** primarily for property protection rather than life safety.	This system covers key areas like storage rooms, archives, electrical rooms, or high-risk areas. It does not typically cover escape routes or individual rooms but ensures that areas with a higher risk of fire are monitored.	Detection coverage in specific high-risk areas (e.g., storage rooms, plant rooms). Suitable for commercial or industrial buildings where property protection is a priority.
5	**L5 – Partially Zoned System**	Designed for buildings that require **partial life protection**, focusing on areas of the building where people are more vulnerable or where the fire risk is higher.	L5 systems focus on specific areas, typically high-risk zones such as plant rooms, server rooms, or specialized areas. These systems may be used in environments where fire protection for property or machinery is more critical than for life protection.	Coverage tailored to specific higher-risk areas. Suitable for certain industrial and commercial buildings.

| 6 | **M – Manual Systems** | A manual fire alarm system relies on occupants to activate the alarm when they spot a fire. | Manual call points (MCPs) are positioned throughout the building, allowing anyone to trigger the alarm manually by pressing a button or pulling a lever. This system is often used in **low-risk** or **smaller** buildings. | Manual activation by building occupants. Typically used in smaller buildings or areas with limited fire risk where automatic detection is unnecessary. Common in small offices, shops, and some residential buildings. |

Grade of Fire Alarm Systems

The grade of the fire alarm system indicates the level of coverage and sophistication. It also refers to the quality of components, wiring, and overall reliability. The description of the various grades and their features are brought out in the table below:

	System Grade	Description	Features
1	**Grade A – Fully Wired System**	In a Grade A system, all components (detectors, control panels, alarm devices, etc.) are **hard-wired** to each other, ensuring high reliability. Grade A systems are typically used in large commercial and industrial buildings or buildings with complex layouts and high-risk environments.	• Fully interconnected components. • Suitable for large or complex premises, offering a high level of redundancy. • Requires professional installation and certification.

2	**Grade B – Wired System with Reduced Coverage**	Similar to Grade A, but with **reduced coverage** or fewer components. Typically used in smaller buildings or premises with lower fire risks.	• Wired system but with less comprehensive coverage than Grade A. • Suitable for smaller or medium-sized buildings.
3	**Grade C – Non-Wired System**	This grade involves using **wireless** or **partially wired** components. These systems are usually suitable for smaller premises or specific applications where cost and flexibility are a priority.	• Wireless or minimal wiring. • Suitable for temporary or smaller-scale installations.
4	**Grade D – Battery Powered Systems**	A Grade D system typically uses **battery-powered smoke detectors** or fire alarms, which are often found in **domestic** settings or **low-risk commercial** premises.	• Battery-operated components. • Often used for domestic or residential fire detection. • May include smoke detectors and manual call points.
5	**Grade E – Non-Interlinked Systems**	This is the lowest level of system, where **smoke detectors** or alarms are **not interlinked**. These systems are often found in **low-risk domestic settings** where only basic protection is required.	• Basic system, typically used in private homes or small buildings. • Alarm activation in one area does not trigger the entire building.

Summary of Fire Alarm System Types The choice of fire alarm system, including both the category and grade, depends on the building type, occupancy, fire risk assessment, and the specific regulatory requirements. For example, a Category L1 system might be required for a high-occupancy public building to provide full coverage, while a Category M system might be sufficient for a small shop where manual activation is the main method of alert. The grade of the system indicates how sophisticated and reliable the system is, with Grade A being the most robust option for large, high-risk environments.

The tables below summarize the scope of the various categories and grades of fire alarm systems.

Category	Description
L1	Full coverage with detectors in all rooms and escape routes.
L2	Zoned coverage with detectors in high-risk areas and escape routes.
L3	Focus on escape routes to ensure safe evacuation.
L4	Property protection in key high-risk areas.
L5	Partial life protection with coverage in specific high-risk zones.

Grade	Description
A	Fully wired system with interlinked components.
B	Wired system with reduced coverage or fewer components.
C	Wireless or partially wired system.
D	Battery-powered system, typically for residential or low-risk premises.
E	Non-interlinked system, generally for smaller or lower-risk buildings.

Chapter 10

Building Codes and Regulations

Fire safety building codes and regulations in the UK are governed by various laws, guidance documents, and standards that aim to reduce the risk of fire, ensure the safety of building occupants, and protect property. These regulations cover the construction, design, and maintenance of buildings, as well as ongoing fire safety management.

Legislation Governing Fire Safety

The Regulatory Reform (Fire Safety) Order 2005

The Regulatory Reform (Fire Safety) Order 2005 (commonly referred to as the Fire Safety Order or RRO) is the cornerstone of fire safety legislation in England and Wales. It came into force on 1st October 2006, consolidating and replacing over 70 pieces of earlier legislation, including the Fire Precautions Act 1971. The aim of the Order is to simplify fire safety law and place the emphasis on fire prevention in all non-domestic premises, which include:

- Workplaces (offices, factories, warehouses)
- Commercial buildings (shops, hotels, restaurants)
- Public buildings (schools, hospitals, theatres, and care homes)
- Common areas of multi-occupied residential buildings (such as lobbies, stairwells, and corridors)

Responsible Person

Under the Order, the primary duty for fire safety lies with the "responsible person", a legal designation that typically refers to:

- The employer, in workplaces
- The person who has control over the premises (e.g., building managers or facilities managers)
- The owner or landlord of the building, in some contexts

In cases where more than one responsible person exists (such as in shared premises), all parties are required to cooperate and coordinate to ensure comprehensive fire safety coverage.

Duties and Responsibilities

The Fire Safety Order is risk-based and places a proactive legal duty on the responsible person to take general fire precautions and ensure the safety of employees and others who may be affected by fire. Key responsibilities include:

- **Fire Risk Assessment**: The responsible person must carry out a suitable and sufficient fire risk assessment, identifying fire hazards, evaluating the risk to people, and determining what measures are needed to ensure safety. The assessment must be reviewed regularly and updated as necessary, especially when significant changes occur (e.g., building layout, use, or occupancy).
- **Preventative and Protective Measures**: The responsible person must implement appropriate fire prevention measures to reduce the likelihood of fire, and protective measures to mitigate the impact should a fire occur. This includes:
    - Eliminating or reducing fire hazards (e.g., managing flammable materials)
    - Installing and maintaining fire detection and warning systems
    - Providing suitable fire-fighting equipment (e.g., extinguishers, sprinkler systems)
    - Ensuring that fire exits are sufficient, clearly marked, and unobstructed
    - Maintaining emergency lighting and signage
    - Establishing procedures for evacuation and fire drills
- **Maintenance and Testing**: Fire safety equipment and systems (e.g., alarms, extinguishers, emergency lighting) must be maintained in efficient working order and subject to regular testing and servicing in line with relevant British Standards.
- **Training and Information**: Employees and relevant persons must be provided with adequate fire safety training and information. This includes:
    - Induction training for new staff
    - Refresher training at appropriate intervals
    - Specific training for fire wardens or marshals
    - Clear instruction on evacuation procedures

- **Emergency Planning**: The responsible person must prepare an emergency plan, which should be tailored to the specific risks and layout of the premises. This should include:
  - Procedures for raising the alarm and evacuating the building
  - Identification of escape routes and muster points
  - Responsibilities of fire wardens
  - Coordination with the fire and rescue service

Enforcement and Penalties

The Order is enforced by fire and rescue authorities, who have the power to:
- Conduct inspections and audits
- Serve enforcement, alteration, or prohibition notices
- Initiate prosecutions for non-compliance

Failure to comply with the Fire Safety Order is a criminal offence and can result in:
- Unlimited fines
- Imprisonment for up to two years for serious breaches
- Closure of premises deemed unsafe

High-profile prosecutions in recent years have underscored the seriousness with which compliance is treated. Businesses and organisations are strongly encouraged to adopt a culture of continuous improvement in fire safety.

Interaction with Other Legislation

While the Fire Safety Order is the primary legislation, it operates in conjunction with other legal frameworks, including:

The Health and Safety at Work etc. Act 1974

The Building Regulations 2010 (especially Approved Document B for fire safety in new buildings)

The Fire Safety Act 2021, which amends the RRO to clarify responsibilities in multi-occupied residential buildings following the Grenfell Tower fire

The Building Safety Act 2022, introducing further oversight and accountability for high-risk buildings

Together, these legal instruments form a comprehensive regulatory regime aimed at improving fire safety across the built environment.

Summary

The Regulatory Reform (Fire Safety) Order 2005 represents a modern, risk-based approach to fire safety regulation. By placing the burden of responsibility squarely on those in control of premises, it ensures that fire safety is considered an integral part of building management and workplace operations. Compliance is not just a legal obligation but a fundamental aspect of protecting life, property, and business continuity.

The Fire Safety Act 2021

The **Fire Safety Act 2021** is a landmark piece of legislation in the United Kingdom, introduced in response to the **Grenfell Tower fire of 14 June 2017**, in which 72 lives were tragically lost. The disaster revealed serious flaws and ambiguities in the fire safety regulatory framework, particularly concerning **high-rise and multi-occupancy residential buildings**. In response, the government launched a series of inquiries and reforms aimed at improving fire safety. One of the earliest legislative responses was the enactment of the Fire Safety Act 2021.

The Act received **Royal Assent on 29 April 2021** and came into force on **16 May 2022**. It applies to **England and Wales** and makes significant amendments to the **Regulatory Reform (Fire Safety) Order 2005 (FSO)**.

Purpose and Objectives

The primary purpose of the Fire Safety Act 2021 is to:

- **Clarify the scope** of the Fire Safety Order 2005
- Extend fire safety responsibilities to include **external wall systems**, **balconies**, and **flat entrance doors**
- Address enforcement gaps and ensure that **building owners and managers** can be held accountable for a broader range of fire safety risks

This clarity is particularly vital for **responsible persons**, those in charge of managing fire safety in buildings, who previously operated under uncertain interpretations of the FSO.

Key Provisions of the Act

1. *Clarification of the Fire Safety Order's Scope*

The Fire Safety Act explicitly states that the Fire Safety Order applies to:
- The structure and external walls of a building, including anything attached to the exterior of those walls, such as cladding systems, insulation, and fixings

- Balconies and windows
- Flat entrance doors, including the doors between individual flats and shared common areas like corridors and stairwells

Prior to the Act, there was legal uncertainty as to whether these parts of a building fell within the scope of the FSO. The Act removes that ambiguity by confirming that these elements **must be included** in fire risk assessments.

2. *Inclusion in Fire Risk Assessments*

Responsible persons are now **legally required** to consider the **fire safety of external wall systems and entrance doors** when conducting or updating their fire risk assessments.

- Fire Risk Assessments must take a holistic approach, encompassing external fire spread risk (particularly via combustible cladding or balconies) and the integrity of compartmentation at the flat entrance level.
- This shift is particularly significant in high-rise and high-risk buildings, where the spread of fire via the external envelope has proven catastrophic, as evidenced at Grenfell.

3. *Strengthening of Enforcement Powers*

Fire and rescue authorities (as enforcing bodies under the FSO) now have **clearer legal authority** to enforce fire safety standards concerning:

- External wall systems
- Flat entrance doors
- Any relevant common parts

They can issue enforcement, alteration, or prohibition notices where there are deemed to be significant fire risks relating to these elements. This change enhances **regulatory oversight** and accountability for building owners and managers.

4. *Application to All Multi-Occupancy Residential Buildings*

The Fire Safety Act applies to **all buildings containing two or more sets of domestic premises** that share common areas, this includes:

- Purpose-built blocks of flats
- Converted houses with multiple flats
- Mixed-use buildings with commercial premises and residential units above

It is important to note that the Act does not impose new duties per se but clarifies existing responsibilities under the FSO to ensure consistent and robust fire risk management across all residential property types.

Implementation and Supporting Measures

The Fire Safety Act forms part of a **broader package of fire safety reforms**, which include:

- The **Building Safety Act 2022**, which introduces new obligations for "higher-risk buildings" (e.g. residential buildings over 18m in height or with 7 or more storeys)
- The **Fire Safety (England) Regulations 2022**, which provide detailed requirements for the ongoing management of fire safety in residential buildings, including:
  - Monthly checks on fire doors
  - Provision of fire safety instructions to residents
  - Installation and maintenance of way finding signage
  - Provision of building plans and external wall system information to the local fire and rescue service

These regulations were developed to **operationalise** the principles of the Fire Safety Act and provide practical mechanisms for compliance.

Industry Impact

The Fire Safety Act 2021 has had a significant impact on the **built environment and property management sectors**, especially for:

- Building owners and freeholders, who must now commission more comprehensive and sometimes more technical fire risk assessments
- Fire risk assessors, who must be competent to evaluate complex external wall systems and understand modern cladding technologies
- Surveyors and construction professionals, who are often called upon to advise on remediation or compliance strategies
- Leaseholders, many of whom have been caught in the financial and legal implications of buildings found to have non-compliant cladding or fire doors

Challenges and Controversies

The implementation of the Fire Safety Act has not been without controversy. Key challenges include:

- Shortage of qualified fire risk assessors with expertise in external wall systems
- Remediation costs falling on leaseholders, often leaving them trapped in unsellable homes
- Backlogs in fire risk assessments and EWS1 forms for mortgage approvals
- Inconsistency in enforcement between fire and rescue authorities

These challenges have led to ongoing debates around funding, fairness, and the capacity of the industry to respond to increased demands.

Conclusion

The Fire Safety Act 2021 represents a pivotal development in the UK's fire safety regulatory framework. By clarifying the responsibilities of building owners and managers regarding external wall systems and flat entrance doors, it ensures that fire safety risk assessments are comprehensive and reflective of real-world risks. Although not a solution to all the issues exposed by Grenfell, it is a foundational step in a wider programme of reform designed to rebuild trust, enhance safety, and protect lives in the built environment.

The Fire Safety (England) Regulations 2022

The Fire Safety (England) Regulations 2022 represent a major step forward in the UK's post-Grenfell fire safety reform programme. They were introduced under the powers of the Fire Safety Act 2021 and came into effect on 23 January 2023, placing new legal duties on the responsible person (as defined in the Regulatory Reform (Fire Safety) Order 2005) to implement enhanced fire safety measures in residential buildings, particularly those deemed high-rise or higher risk.

These Regulations apply only in England, and form part of the government's broader commitment to improving fire safety standards in multi-occupancy residential buildings, especially in light of lessons learned from the Grenfell Tower tragedy.

Purpose and Intent

The primary aims of the Fire Safety (England) Regulations 2022 are to:
- Strengthen fire safety management and oversight in residential buildings
- Ensure that residents are better informed about fire safety measures
- Improve preparedness for emergency response, particularly for fire and rescue services
- Reduce the risk of fire spread and support safe evacuation or rescue

Scope and Application

The Regulations apply to:

- All multi-occupied residential buildings with common parts (e.g. hallways, stairwells)
- Additional duties for buildings over 11 metres in height
- Further enhanced requirements for high-rise residential buildings (i.e. buildings at least 18 metres tall or at least 7 storeys)

The level of obligation increases with the height and complexity of the building, recognising the elevated risk and challenges associated with high-rise fire incidents.

Key Duties Imposed by the Regulations

The Regulations introduce **three tiers of duties**, based on the building's height and risk profile.

1. *All Multi-Occupied Residential Buildings (Any Height)*

For all such buildings, the **responsible person** must:

- Provide fire safety instructions to all residents:
    - Explaining how to report a fire
    - Outlining the building's evacuation strategy (e.g., "stay put" or "simultaneous evacuation")
    - Providing guidance on what to do in the event of a fire
- **Provide information about fire doors**:
    - Emphasising the importance of keeping fire doors closed
    - Highlighting how fire doors help prevent smoke and flame spread
    - Detailing how to report damaged or faulty doors

This ensures that all residents, regardless of the building's height, are better informed and empowered to act appropriately in the event of a fire.

2. *Buildings Over 11 Metres in Height*

In buildings exceeding 11 metres in height, the responsible person must additionally:

- Undertake quarterly checks of all fire doors in communal areas
- Annually inspect the entrance doors to individual flats, where possible

These inspections are crucial to maintaining compartmentation and ensuring that fire doors function as intended to contain fire and smoke, protecting escape routes and limiting fire spread.

3. *High-Rise Residential Buildings (18 Metres or 7+ Storeys)*

For the highest-risk buildings, the Regulations introduce a comprehensive suite of duties, primarily focused on supporting fire and rescue services in carrying out effective firefighting and rescue operations:

- **Provide building information to fire and rescue services, including:**
  - Electronic copies of floor plans and building layout information
  - Location of key fire safety systems (e.g., fire mains, smoke control systems)
  - Details of fire-fighting lifts and evacuation lifts
- **Install and maintain way finding signage:**
  - Floor identification and flat numbers must be clearly visible in low-light or smoke-filled conditions
  - Signage should be visible from the staircase and common corridors to assist both residents and responders
- **Install secure information boxes (Premises Information Boxes):**
  - These must be readily accessible to fire and rescue services
  - Boxes should contain key information, including:
    - Building plans
    - Contact details of the responsible person
    - Fire safety strategy and evacuation arrangements
- **Routine checks of essential fire safety systems**, such as:
  - Fire detection and alarm systems
  - Emergency lighting
  - Firefighting equipment
  - Smoke control systems
  - Fire door self-closing devices

The responsible person must record the results of these checks and be prepared to share them with the local fire and rescue service upon request.

Compliance and Enforcement

- **Fire and rescue services** are the enforcing authority for the Fire Safety (England) Regulations.

- Failure to comply can result in **enforcement notices**, **prohibition notices**, or **prosecution** under the Regulatory Reform (Fire Safety) Order 2005.
- Building owners and managing agents must ensure that **competent persons** are engaged to carry out inspections, risk assessments, and system maintenance.

Integration with Other Legislation

The Fire Safety (England) Regulations 2022 operate in conjunction with:
- **The Fire Safety Act 2021**, which clarifies that the external wall systems and flat entrance doors must be included in fire risk assessments
- **The Building Safety Act 2022**, which introduces a new regulatory framework for high-risk buildings, including the **Accountable Person** and **Building Safety Regulator**
- **BS 9991** and **BS 9999**, which provide best-practice fire safety design and management guidance

Together, these laws and standards form a **comprehensive and layered approach** to fire safety, targeting both building design and operational management.

Impact and Implications

The introduction of the Fire Safety (England) Regulations 2022 has had far-reaching effects on:
- **Building owners and freeholders** – who must implement new inspection, maintenance, and documentation practices
- **Property managers and agents** – who face increased operational responsibility for **compliance**
- **Fire safety professionals and assessors** – who must support clients in meeting new regulatory demands
- **Residents** – who are now more informed and engaged in the fire safety measures of their homes

These Regulations mark a shift from a reactive to a proactive fire safety culture, with a focus on ongoing oversight, transparency, and resident communication.

Conclusion

The Fire Safety (England) Regulations 2022 represent a critical development in the UK's evolving fire safety framework. They introduce specific, actionable requirements designed to mitigate fire risks, enhance evacuation preparedness, and support effective emergency response, particularly in high-rise and complex residential buildings. As part of a broader legislative transformation, these Regulations reinforce the need for continuous fire safety management and a well-informed residential community, forming an essential layer in protecting lives and property in the post-Grenfell era.

Building Regulations 2010 (Part B) – Fire Safety Requirements

The Building Regulations 2010 (Part B) is divided into two volumes:
- **Volume 1**: Dwellings (houses, flats, etc.)
- **Volume 2**: Buildings other than dwellings (offices, factories, etc.)

Key aspects of fire safety in the Building Regulations include:

B1: Means of warning and escape

This section focuses on ensuring that adequate fire detection and alarm systems are in place, as well as providing safe and efficient means of escape for occupants. The routes should be protected from fire and smoke, leading to a place of safety.

B2: Internal fire spread (linings)

This deals with the fire resistance of internal wall and ceiling linings, such as plasterboard and paint, to reduce the spread of flames and smoke within a building.

B3: Internal fire spread (structure)

This section aims to prevent the spread of fire by requiring structural elements (walls, floors, etc.) to have adequate fire resistance. It includes provisions for compartmentation—dividing the building into fire-resisting sections to prevent the spread of fire.

B4: External fire spread

This aspect addresses the spread of fire on external walls and roofs. It became particularly important after the Grenfell Tower tragedy, focusing on the use of non-combustible cladding materials and fire safety barriers to prevent fire from spreading to adjacent buildings.

B5: Access and facilities for the fire service

This ensures that buildings are designed to allow the fire and rescue services to access them quickly in an emergency. It includes provisions for fire hydrants, firefighting shafts, and vehicle access.

Fire Risk Assessments

Under the Regulatory Reform (Fire Safety) Order 2005, responsible persons must conduct regular fire risk assessments to identify potential hazards and ensure appropriate fire safety measures are in place. The steps involved in a fire risk assessment include:

1. **Identifying fire hazards**:
   Identifying sources of ignition, fuel, and oxygen.
2. **Identifying people at risk**
   Considering employees, visitors, contractors, and anyone who may be more vulnerable (e.g., disabled or elderly individuals).
3. **Evaluating and reducing risks**
   Implementing measures to reduce the likelihood of fire, such as removing ignition sources, improving fire detection, and ensuring evacuation routes are clear and accessible.
4. **Recording findings**
   Keeping detailed records of the risk assessment and fire safety measures.
5. **Reviewing the assessment**
   Regularly reviewing the fire risk assessment to ensure that it remains up to date.

High-Rise Buildings and the Aftermath of Grenfell

The Grenfell Tower fire in 2017 led to a comprehensive review of fire safety in the UK, particularly regarding high-rise residential buildings. This resulted in:

The Hackitt Review

An independent review of building regulations and fire safety led by Dame Judith Hackitt. The review identified systemic failings in the construction industry and recommended a new regulatory framework to improve fire safety, particularly in high-rise buildings.

Ban on Combustible Cladding

Following the Grenfell disaster, the UK government banned the use of combustible materials in the external walls of new high-rise buildings over 18 meters in height. This will be reduced 11 meters in upcoming editions of ADB.

Fire Safety Act 2021

This legislation was introduced to address the specific risks posed by the external wall systems (including cladding) of multi-occupancy residential buildings. It clarified the responsibilities for the safety of external walls and front doors of individual flats, making building owners responsible for assessing the safety of cladding and other fire risks.

Fire Safety Management in Residential and Commercial Buildings

In addition to construction and structural regulations, ongoing fire safety management is crucial. This includes:

Emergency Lighting

Buildings must have appropriate emergency lighting to ensure that escape routes are visible in the event of power failure during a fire.

Fire Alarm and Detection Systems

These should be appropriate for the size and layout of the building, with systems regularly maintained and tested.

Fire Doors

Fire doors must be installed in key locations to prevent the spread of fire and smoke. They should be kept closed or fitted with automatic closers and maintained regularly.

Sprinkler Systems

While not mandatory in all buildings, the installation of sprinkler systems is highly recommended, particularly in high-risk buildings. Recent legislation requires sprinklers to be installed in all new high-rise residential buildings over 11 meters in height.

Maintenance of Fire Safety Equipment

All fire safety systems, including alarms, extinguishers, emergency lighting, and sprinkler systems, must be maintained and tested regularly to ensure they are fully operational in the event of a fire.

Enforcement and Penalties

Fire service enforcement in the UK is a key aspect of ensuring that fire safety regulations are properly followed across various types of buildings. The enforcement process is primarily handled by local Fire and Rescue Authorities (FRAs), which have the responsibility and authority to inspect buildings, assess fire safety measures, and take action if there are any breaches of fire safety laws. Here's a more detailed explanation of how fire service enforcement works:

Powers and Duties of Fire and Rescue Authorities (FRAs)

The Regulatory Reform (Fire Safety) Order 2005 gives fire and rescue services wide-ranging powers to enforce fire safety in non-domestic premises (including the common areas of residential buildings such as flats). Their main roles include:

Inspection of Premises

Fire officers have the legal right to inspect buildings at any reasonable time without prior notice to assess compliance with fire safety regulations. Inspections may be routine or in response to complaints or concerns. Fire safety inspectors typically look for:

- Adequate fire risk assessments.
- Fire alarm and detection systems.
- Firefighting equipment (e.g., extinguishers).
- Fire-resistant construction and materials.
- Properly maintained fire doors and emergency lighting.
- Clear and safe evacuation routes.

Issuing Notices

If a fire safety inspector identifies deficiencies in fire safety during an inspection, they can issue formal notices to rectify the situation. There are several types of notices:

- **Alterations Notice**

Issued if a building presents a high fire risk, or if there are plans for alterations that might impact fire safety. The responsible person is required to notify the authority before making any significant changes.

- **Enforcement Notice**

Issued when there are serious breaches of fire safety regulations. The responsible person is given a set period to correct the issues. The notice will specify what needs to be done to comply with the law and by when.

- **Prohibition Notice**

This is issued if the building or part of the building is deemed to pose a serious risk to occupants' safety. It can lead to the closure of premises or prohibit the use of certain parts until fire safety improvements are made.

Monitoring Compliance

Once a notice has been issued, the fire service will often revisit the premises to ensure that the required actions have been taken within the set time frame. Non-compliance can lead to further legal action.

Penalties for Non-Compliance

Failure to comply with fire safety regulations can result in severe penalties. Fire and rescue services can take legal action against the responsible person(s) in charge of a building if they fail to comply with fire safety rules or notices. Penalties include:

Fines

Minor offences may result in a fine, but fines can also be substantial for more serious breaches. There is no fixed maximum fine for offences under the Regulatory Reform (Fire Safety) Order 2005, and courts can impose large fines depending on the severity of the breach.

Imprisonment

For more severe violations, particularly those that put lives at serious risk, the responsible person can be sentenced to imprisonment. In some cases, individuals have been sentenced to up to two years in prison for serious fire safety failures.

Court Orders

In extreme cases where life safety is severely compromised, courts may issue orders to close the building until adequate fire safety measures are put in place.

Risk-Based Approach to Enforcement

FRAs often use a risk-based approach to prioritize their enforcement activities. This means that they focus their attention on premises that present the highest risk to life. This could include:

High-rise residential buildings

Particularly after the Grenfell Tower fire, high-rise buildings are now subject to closer scrutiny, especially those with external cladding and complex evacuation plans.

Hospitals, care homes, and schools

These premises house vulnerable individuals who may have limited mobility or may not be able to evacuate quickly in the event of a fire, requiring stringent fire safety measures.

Workplaces and public buildings

Offices, factories, warehouses, and public entertainment venues where large numbers of people congregate are also given priority.

Enforcement in High-Risk Buildings: Post-Grenfell Changes

The Grenfell Tower fire in 2017 significantly impacted fire service enforcement, especially regarding high-rise residential buildings and those with cladding systems. Post-Grenfell measures include:

Cladding Investigations

After Grenfell, fire services were heavily involved in assessing buildings with potentially dangerous cladding (e.g., Aluminium Composite Material or ACM) to ensure that it was replaced with safer materials.

Waking Watch Schemes

In some cases, enforcement actions led to the implementation of temporary waking watch schemes, where trained staff monitor buildings 24/7 to raise the alarm in the event of a fire until permanent fire safety systems are installed.

Fire Safety Remediation

Fire services now collaborate closely with local authorities, building owners, and housing associations to ensure that high-risk residential buildings are remediated to comply with fire safety standards. This includes changes to the building's cladding, fire doors, alarms, and evacuation protocols.

Joint Working with Other Enforcement Bodies

Fire and rescue services often work in collaboration with other regulatory bodies to enforce fire safety laws, including:

Local Authorities

They are responsible for certain fire safety aspects, particularly in housing. For example, if a residential property presents hazards under the Housing Health and Safety Rating System (HHSRS), the local authority can take action to improve conditions in private rented housing.

Health and Safety Executive (HSE)

Fire services sometimes collaborate with the HSE, particularly in workplaces, to ensure that fire safety measures are part of a broader health and safety plan.

Building Control Bodies

During the construction or alteration of buildings, fire services often work with local authority building control officers or approved inspectors to ensure compliance with fire safety regulations in new building projects.

Public Accountability and Transparency

Fire and rescue services are expected to operate transparently, making their enforcement activity public. The Fire Safety Order 2005 encourages FRAs to publish enforcement actions taken against buildings that pose fire safety risks. This public accountability serves as a deterrent for non-compliance and encourages building owners to maintain high fire safety standards.

Supporting Compliance: Advice and Guidance

In addition to enforcement, fire services play a key role in supporting building owners, managers, and employers by providing fire safety advice. They often run community and business outreach programs to raise awareness about fire safety requirements, educate people on how to carry out fire risk assessments, and promote best practices in fire prevention.

Summary of Fire Service Enforcement

Fire service enforcement in the UK is robust and multifaceted, focusing on ensuring compliance with fire safety regulations to protect lives and property. Fire authorities use a risk-based approach, targeting inspections and enforcement actions at high-risk buildings such as high-rise residential properties, hospitals, and schools. Failure to comply with fire safety regulations can result in enforcement notices, hefty fines, and even imprisonment. The role of fire services has expanded significantly post-Grenfell, with stricter regulations around high-rise buildings, cladding safety, and joint work with other enforcement bodies.

The overall goal of fire service enforcement is to create safer buildings through a combination of inspection, advice, and legal action where necessary.

Chapter 11

Fire Safety Engineering Design

Fire engineering design of buildings in the UK is a critical discipline that ensures buildings are designed, constructed, and maintained to minimize the risk of fire and protect both occupants and property. The practice encompasses various aspects such as fire safety design, egress (escape) routes, evacuation planning, the use of fire barriers, and compartmentalization. Here's a comprehensive account of these key elements:

Fire Engineering Design in the UK

Fire engineering in the UK goes beyond prescriptive regulations to incorporate a performance-based approach that evaluates how a building will perform in a fire. It involves the application of scientific principles to predict fire behaviour, evaluate building designs, and ensure that safety objectives are met.

STANDARDS AND REGULATIONS

The primary guidance for fire engineering design comes from:

- **Building Regulations 2010 (Approved Document B)**: This provides the legal framework for fire safety in buildings, detailing prescriptive measures that buildings must meet. However, fire engineering allows for alternative solutions to these prescriptive measures, as long as the design meets equivalent safety standards.

- **BS 9999: Code of Practice for Fire Safety in the Design, Management, and Use of Buildings**: A performance-based fire safety standard that offers flexibility in fire design by allowing for the variation of fire protection measures depending on the building's use, occupancy, and fire risks.

- **BS 7974: Application of Fire Safety Engineering Principles to the Design of Buildings**: This British Standard provides a framework for the development of fire safety engineering strategies. It includes guidance on the assessment of fire risks, fire spread, and structural fire protection.

Fire engineering can involve designing systems that control fire spread, limit smoke movement, maintain structural integrity during a fire, and ensure safe evacuation.

Designing for Fire Safety in Buildings

Fire safety design begins in the early stages of building design and focuses on ensuring that the building can prevent the outbreak of fire, slow its spread, and facilitate the safe evacuation of occupants. It involves several key areas:

Fire Prevention and Detection Systems

Effective fire safety design involves incorporating active and passive fire protection systems into the building's design:

- **Active fire protection** includes systems like fire detection and alarm systems, automatic sprinkler systems, and smoke control systems. These systems actively detect fire, suppress it, or control its spread.
- **Passive fire protection** refers to built-in measures that resist fire and heat, such as fire-rated walls, floors, and doors. These elements prevent the spread of fire and smoke by maintaining structural stability.

Materials and Structural Fire Resistance

The use of fire-resistant materials is critical in fire engineering design. These materials are selected based on their ability to resist ignition and withstand high temperatures for extended periods. This includes:

- **Fire-resistant construction materials** for walls, floors, and ceilings, which can contain fire within a compartment.
- **Intumescent coatings** or fireproofing materials that expand when exposed to heat, forming an insulating barrier to protect structural elements like steel beams.

Fire Load Calculation

Fire load is the potential energy available for combustion in a building. A fire engineering design will assess the fire load based on the building's contents, construction materials, and intended use. This helps in determining the level of fire protection required and the expected fire behaviour.

Smoke Ventilation and Control Systems

Smoke is often more dangerous than the fire itself. In high-rise buildings or complex structures, fire engineering design includes smoke control systems, such as:

- **Smoke vents** to allow smoke to escape.
- **Pressurized stairwells** to keep evacuation routes clear of smoke.
- **Smoke curtains** to direct smoke away from escape routes or sensitive areas.

Egress Design and Evacuation Planning

One of the most critical aspects of fire engineering design is ensuring that all building occupants can safely evacuate in the event of a fire. This is achieved through careful egress design and comprehensive evacuation planning.

Means of Escape

The design of **escape routes** ensures that occupants can leave a building quickly and safely in an emergency. Key considerations include:

- **Number and location of exits**: Large or complex buildings require multiple exits to ensure that no single escape route becomes congested or unusable in a fire. These exits must be strategically located to ensure occupants can reach safety from any point in the building.
- **Dimensions of escape routes**: Escape routes must be wide enough to allow for the smooth flow of people. The width of corridors, staircases, and doorways is determined by the number of occupants using them, ensuring they can evacuate without delay.
- **Fire-rated escape routes**: Escape routes need to be protected from fire and smoke, often by enclosing them in fire-resistant materials, such as fire-rated walls and doors, that provide enough time for safe evacuation.

Evacuation Time Analysis

Fire engineers use computational modelling tools to estimate the **Available Safe Egress Time (ASET)** and compare it to the **Required Safe Egress Time (RSET)**. ASET is the time available before conditions become life-threatening, while RSET is the time occupants need to evacuate. The building design ensures that RSET is shorter than ASET, providing enough time for safe evacuation.

Evacuation for Vulnerable Groups

Egress design must account for vulnerable groups, including those with mobility impairments. This may include:

- **Evacuation lifts**: Fire-rated lifts designed to transport people who cannot use staircases.
- **Refuge areas**: Safe areas where disabled or vulnerable occupants can wait for assistance during evacuation.

Evacuation Strategy

Evacuation strategies vary depending on the building type and occupancy. Common strategies include:

- **Simultaneous Evacuation**: In small or simple buildings, all occupants evacuate at once upon the fire alarm.
- **Phased Evacuation**: In large buildings like offices or hospitals, occupants near the fire evacuate first, followed by other areas in stages. This reduces congestion and maintains order.
- **Stay Put Policy**: For high-rise residential buildings, the "stay put" policy suggests that residents stay in their flats unless the fire is in their unit, as buildings are designed with compartmentalization to prevent fire spread.

Use of Fire Barriers and Compartmentalization

Fire Barriers

Fire barriers are designed to prevent the spread of fire and smoke between different parts of a building. They are typically constructed from fire-resistant materials, such as fire-rated walls, floors, ceilings, and doors. Their primary role is to compartmentalize a building into smaller, fire-resisting sections.

Compartmentation

Compartmentation is a fundamental principle of fire engineering that involves dividing a building into separate compartments to limit the spread of fire and smoke. Each compartment is enclosed by fire-resistant construction materials, creating a barrier that can contain a fire for a specific duration, typically 30, 60, 90, or 120 minutes, depending on the fire safety strategy.

Compartmentation in Practice

- **Horizontal and vertical compartmentation**: Fire compartments are typically separated by fire-rated floors (horizontal) and walls (vertical). Vertical compartmentation is particularly critical in multi-story buildings to prevent fire from spreading between floors.
- **Fire-resisting doors**: **Fire doors** are an integral part of compartmentation. They are designed to remain closed during a fire and resist fire for a specified period (e.g., FD30 fire doors resist fire for 30 minutes). These doors help protect escape routes and maintain the integrity of fire compartments.

- **Penetrations and service routes**: Gaps created by services (pipes, cables, ducts) passing between compartments must be sealed with fire-stopping materials to maintain the compartment's fire resistance.

Purpose of Compartmentalization

The primary objectives of compartmentalization are:

- **Containment of fire and smoke**: Limiting the spread of fire to its area of origin slows the progression of the fire, providing more time for evacuation and allowing firefighting teams to control the blaze.
- **Protection of escape routes**: Compartments protect key escape routes (corridors and stairwells) from being compromised by fire or smoke, ensuring occupants can evacuate safely.
- **Property protection**: By containing the fire, compartmentation reduces the risk of extensive property damage.

Fire Safety and Modern Building Design

Fire engineering design also needs to address the unique challenges posed by modern architectural trends, such as:

- **High-rise buildings**: High-rise structures require special consideration for egress routes (e.g., pressurized stairwells), smoke control systems, and the use of non-combustible cladding.
- **Open-plan spaces**: These designs often lack natural compartmentation, so fire engineers must use innovative solutions, such as **water mist systems** or **fire curtains**, to limit fire spread in the absence of traditional walls.
- **Sustainable materials**: Increasing use of eco-friendly materials in construction (e.g., timber) necessitates additional fire engineering considerations, ensuring that these materials meet fire safety requirements without compromising sustainability goals.

Fire engineering design in the UK is a complex, multidisciplinary approach that balances fire safety with modern building design and functionality. It involves a thorough understanding of fire behaviour, structural fire resistance, and human evacuation dynamics. By focusing on passive measures like compartmentation and fire barriers, along with active fire protection systems and tailored evacuation strategies, fire engineering ensures that buildings can provide a safe environment for occupants in the event of a fire. The performance-based nature of fire engineering allows for flexibility in design, while maintaining rigorous safety standards that adapt to the specific challenges of each building type.

Performance Based Building Design

Principles of Performance-Based Design

Performance-based building design (PBD) is a modern approach to building design that focuses on achieving specific performance outcomes rather than adhering strictly to prescriptive codes and standards. This method is particularly relevant in the fields of Fire Safety Engineering, Structural Engineering, and Energy Efficiency, where the complexity of modern buildings and the variety of materials and technologies make prescriptive rules less applicable. When adopting this approach Project Management is essential to organize and link together the different fields of expertise required. PBD involves many disciplines from the construction industry, this chapter gives an overview of PBD then focuses on the role of the fire engineer.

Performance-based design is rooted in achieving specific safety and functional outcomes, allowing designers greater flexibility in the methods and technologies they use to achieve these goals. The key principles include:

Goal-Oriented Approach

The Goal-Oriented Approach is a fundamental aspect of performance-based building design, as it shifts the focus from complying with fixed, prescriptive building codes to achieving specific, measurable performance outcomes. This approach allows for more flexibility and innovation in the design process, as the primary objective is to meet the performance goals that are most critical to the project. The key performance criteria often include occupant safety, property protection, sustainability, and resilience, among others.

Occupant Safety
Prioritizing Life Safety

In traditional building design, safety measures are often designed by prescriptive codes and regulations that specify the number of exits, fire resistance ratings, and ventilation requirements. In a goal-oriented approach, however, the focus is on ensuring that all building occupants can safely evacuate in an emergency, regardless of the methods used to achieve this outcome.

Flexible Solutions

Instead of simply following predefined standards for the safety (e.g. a specific number of fire exits), designers might use advanced evacuation modelling, smoke control systems, or early warning systems that are tailored to the specific layout and use of the building. For example, in large and complex buildings such as airports, sports arenas or pharmaceutical establishments, performance-based design might include phased evacuations or the use of refuge areas.

Risk-Based Assessments

Designers can assess the specific risks associated with the building, such as the likelihood of fire or structural collapse, and implement measures that directly address those risks. This could mean enhancing fire suppression systems in areas with high fire loads or strengthening certain parts of the structure to resist earthquake forces.

Property Protection
Preserving Building Integrity

Property protection in performance-based design is not just about meeting regulatory requirements for fire resistance or load-bearing capacity. It is about ensuring that the building can continue to function or be restored quickly after an incident. This is particularly important for critical infrastructure or facilities where downtime can have significant economic or societal impacts, such as hospitals or data centres.

Tailored Fire and Structural Engineering Solutions

A goal-oriented design for property protection might focus on ensuring that key parts of the building (e.g. structural components, IT infrastructure) are shielded from damage in the event of a fire or flood. This could involve installing fire-resisting barriers and/or redundant power systems.

Business Continuity

The design might also prioritise business continuity by ensuring that even if part of the building is damaged, other parts can remain operational. For example. In a performance-based fire strategy, designers might ensure that key operational areas are protected through compartmentation, advanced fire suppression, or alternative power systems.

Sustainability
Environmental Performance Goals

Performance-based design offers considerable flexibility in meeting sustainability goals. Rather than adhering to prescriptive standards like minimum insulation levels or window-to-wall ratios, designers can use performance metrics to optimize the buildings energy use, water consumption, and material selection.

Energy Efficiency

For example, in a performance-based approach, energy models are used to predict the buildings energy consumption under real-world conditions. The goal might be to minimise energy usage or achieve net-zero energy performance. Designers can then choose strategies that best meet that goal, such as passive solar design, high-efficiency HVAC systems, or building-integrated renewable energy sources.

Sustainable Materials and Life Cycle Impact

Performance-based designs also allow for greater consideration of the environmental impacts of materials over the life cycle of the building. Instead of simply meeting a minimum standard for sustainable material use, designers might select materials based on their carbon footprint, durability, and ease of recycling, thus aligning with broader sustainability goals.

Resilience to Natural Hazards

Earthquake, Flood and Wind Resistance

In regions prone to natural disasters, performance-based design can be used to enhance the resilience of buildings to earthquakes, floods, or high winds. The design process might involve advanced modelling to predict how the building will perform under extreme loads and then implementing systems that mitigate damage.

Design for Recovery

Resilience goals in a performance-based approach are not only about surviving a disaster but also about enabling the building to recover quickly. This could involve the use of modular components that can be replaced quickly, resilient foundations that minimize earthquake damage, or flood defences that protect critical systems and allow for quick drainage and clean-up after an event.

Cost-Effectiveness

Optimised Solutions

By focusing on performance outcomes rather than prescriptive standards, performance-based design allows for more cost-effective solutions. Designers can allocate resources where they are needed most, rather than overspending on safety or sustainability measures in areas where they might not provide as much value.

Long-Term Savings

Performance-based design can result in buildings that are more cost effective over their lifespan by reducing energy consumption, minimizing maintenance needs, and lowering the risk of damage from fires or natural disasters. While the initial investment might be higher, the long-term operational savings and reduced risk of downtime or rebuilding can make the approach more financially attractive.

Flexibility for Complex or Unique Buildings

Flexible Design for Unique and Complex Buildings

Many modern buildings have complex designs or serve unique purposes (e.g. Very tall buildings, Stadiums, Airports or Data Centres) that do not fit neatly within the prescriptive frameworks of traditional building regulations. A performance-based approach allows for tailored solutions that fit the specific needs of these buildings.

Innovative Technologies

The approach allows designers to incorporate modern technologies, such as intelligent building management systems, green roofs, or advanced structural materials (e.g. Cross laminated timber or carbon fibre composites), which may not be adequately addressed by existing regulations.

Summary of the Goal-Oriented Approach in Performance-Based Design

The goal-oriented approach in performance-based design empowers architects and engineers to focus on achieving defined performance criteria, such as life safety, structural integrity, energy efficiency, or resilience, without being bound by prescriptive regulations. This results in more flexible, innovative, and tailored solutions that are better suited to the specific needs of modern, complex buildings. By prioritizing outcomes, designers can ensure that the building meets its functional and safety objectives in a cost effective and sustainable manner.

Use of Engineering Judgement

Engineering judgement is the application of professional expertise and decision making to design solutions that meet specific performance objectives while considering real-world constraints and conditions. In performance-based building design, this involves going beyond strict adherence to predefined rules and applying thinking and specialized knowledge to achieve desired outcomes.

Engineering Judgement -
Tailored Solutions to Complex Challenges
Customised Fire Safety Solutions

Instead of adhering to standard fire safety regulations, engineers might use judgement to develop bespoke fire safety systems. For example, in a high-rise building, traditional fire exits might not be practical for certain floors, so engineers could design specialized smoke control systems for refuge areas based on advanced fire modelling to ensure occupant safety.

Adapting Structural Systems

Engineering judgement allows designers to make decisions that optimize structural systems based on performance goals. For example, in earthquake prone regions, engineers can use advanced modelling to determine the optimal use of base isolators or dampers to reduce seismic forces on the building.

Incorporating New and Emerging Technologies - Adoption of Innovative Materials

Engineers can evaluate the performance of new building materials (e.g. cross-laminated timber, high performance concrete, or carbon fibre composites) and integrate them into the design, even if these materials are not explicitly covered by existing codes and regulations.

Smart Systems Integration

Engineering judgement allows the integration of smart building technologies such as sensors, automated fire suppression systems, or real time structural health monitoring. For example, a smart HVAC system can be linked to fire detection systems to automatically adjust airflow and aid in smoke control during a fire.

Deviating from Prescriptive Regulations When Necessary
Justification Through Performance Outcomes

When prescriptive regulations may not suit the specific needs of a project, engineers can use modelling, simulations, and expert knowledge to justify deviations from standard requirements. For instance, if local fire regulations require certain escape routes but the building layout makes this impractical, engineers can use evacuation modelling to prove that alternative routes meet or exceed safety objectives.

Flexibility and Innovation in Performance-Based Design

Performance-Based design promotes flexibility and innovation by allowing designers to pursue creative and non-traditional methods to meet building performance goals. This flexibility encourages the use of modern technologies, construction techniques, and materials that improve building functionality, safety, and sustainability.

Flexibility and Innovation: Use of Advanced Materials

Cross-Laminated Timber (CLT)

Traditionally, wood has been restricted in certain building types due to fire risks. However, performance-based designs using CLT allow for its use in mid and high-rise buildings by addressing fire performance through engineering techniques like charring rate calculations, compartmentation, and sprinkler systems. This flexibility fosters sustainable design by leveraging renewable materials.

Innovative Insulation and Cladding Systems

In performance-based energy designs, buildings can incorporate advanced insulation systems, such as aerogels or phase-change materials, to improve thermal performance without needing to adhere to conventional insulation thickness standards.

Adoption of New Construction Techniques

Modular Construction

Performance-based design allows flexibility in construction techniques. Modular construction, where prefabricated units are assembled on-site can be employed to reduce construction time and improve quality control. Performance assessments can focus on how these modules perform in terms of fire safety, structural integrity, and sustainability.

3D Printing and Off-Site Manufacturing

With advances in 3D printing for construction, performance-based design enables the evaluation of these innovative building processes by focusing on the structural and material performance of 3D printed components, ensuring they meet safety and durability goals.

Incorporation of Smart Building Technologies

Intelligent Fire Detection and Suppression

Performance-based designs can incorporate smart fire detection systems that use IoT sensors to monitor fire conditions in real time, automatically activating suppression systems and adjusting ventilation to control smoke. This technology driven approach goes beyond traditional fire alarms and sprinklers by creating a dynamic, adaptable fire safety system.

Building Management Systems (BMS)

Advanced BMS technologies can monitor a buildings performance in terms of energy use, occupant comfort, air quality, and safety in real time. Performance-based designs allow for customized BMS solutions that adjust building systems automatically to optimize performance, reduce energy consumption, and respond to emergencies.

Energy and Sustainability Innovations

Zero-Net Energy Buildings

In performance-based energy designs, engineers can leverage a range of technologies to create zero-net energy buildings, buildings that generate as much energy as they consume. This flexibility allows the integration of solar panels, geothermal heating, or building integrated photovoltaics (BIPV), without being constrained by prescriptive standards for energy efficiency.

Vertical Green Spaces and Biophilic Design

Performance-Based sustainability strategies enable the use of vertical gardens or green roofs that reduce heat island effects, improve air quality, and enhance energy efficiency, while also providing aesthetic and biophilic benefits for occupants.

Risk-Informed Design in Performance-Based Building Design

Comprehensive Risk Assessment

Identification of Hazards

In a risk-informed approach, engineers begin by identifying all potential hazards that the building might face. This includes fires, structural overloads, natural disasters (e.g. earthquakes, floods, hurricanes), or human-caused events (e.g. explosions, vandalism).

Quantifying Risks

Once hazards are identified, engineers assess the likelihood and severity of each hazard. This allows for a more precise allocation of resources, focusing on the risks that are most probable or could have the most severe consequences. For example, in a region prone to earthquakes, the design might prioritise seismic resilience over fire protection

Tailored Mitigation Strategies
Seismic Design

Although rare, earthquakes do happen in the UK. In areas prone to earthquakes, performance-based design can integrate advanced structural elements like base isolators, which absorb seismic energy, or damping systems that dissipate vibrations. These solutions are tailored to the risk profile of the buildings location and exceed the general requirements of traditional seismic codes.

Flood Resilient Design

For buildings in flood prone areas, risk informed design could include elevated foundations, flood barriers, or waterproofing materials to protect critical building systems. Rather than applying one-size-fits-all flood standards, engineers can use site specific data to design defences that are proportionate to the expected flood risk.

Performance-Based Fire Safety
Smoke Control in Complex Buildings

In performance-based fire safety, engineers may implement smoke control strategies based on risk assessments that account for building layout, ventilation systems, and potential fire loads, in large, complex structures (e.g. airports or shopping Centres) engineers might model smoke behaviour to ensure that evacuation routes remain clear during a fire, using active systems like pressurized stairwells and corridors or automatic smoke vents.

Fire Resistance Tailored to Risk

Instead of applying uniform fire resistance ratings across all building components, risk-informed design focuses on protecting the most critical areas. For example, areas housing high-value equipment or essential utilities may require greater fire resistance or redundant fire suppression systems to ensure business continuity.

Redundancy and Resilience - Redundant Safety Systems

Risk Informed design often includes redundancy for critical systems. In a data centre, for example, both fire suppression systems and backup power supplies might be designed with multiple layers of redundancy to ensure continued operation in case of failure.

Resilience to Multiple Hazards

Performance-based designs can account for the simultaneous impact of multiple hazards. For instance, in a region prone both to hurricanes and floods, the building may need to be designed to withstand high winds while also being protected against storm surges.

Scenario Planning and Simulation
Predictive Modelling

Risk informed design often involves the use of simulations to predict how the building will perform under various hazard scenarios. For example, computational models can simulate the impact of an earthquake, fire, or flood on the buildings structure and systems, allowing engineers to make informed decisions about where to reinforce or add protective measures.

Worst Case Scenarios

Engineers may also develop worst case scenarios, such as a complete failure of fire suppression systems or a direct hit from an earthquake, to assess the buildings resilience and ensure that even under extreme conditions, the building can still meet minimum safety standards.

The Role of Fire Safety Engineering

In fire safety engineering, performance-based design is a sophisticated approach that focuses on achieving specific fire safety outcomes based on the unique characteristics of a building, its occupancy, and its intended use. Instead of following a one-size-fits-all set of prescriptive codes, performance-based design allows for more flexibility and innovation in developing fire safety strategies that are customized for each building. This approach is especially useful in complex structures, high-occupancy environments, or buildings with unique architectural features where prescriptive codes may not be adequate or feasible.

By using fire safety engineering principles, designers can evaluate how a building will behave during a fire and ensure that it meets performance goals such as life safety, property protection, and continuity operations. The design is typically tested under a variety of fire scenarios using advanced modelling and simulations.

Fire Safety Performance-Based Design
Fire Scenarios
In performance-based fire safety design

One of the first steps is to define and model a range of fire scenarios. These scenarios help predict how fire could start, spread, and affect the building and its occupants. Scenarios can range from normal fire loads to worst-case events, such as a large fire occurring in a heavily populated area of the building or in an area with significant amounts of combustible materials.

Plausible Fire Scenarios

Engineers develop fire scenarios based on the building's usage, fire load (i.e., the amount of combustible material), ignition sources, and occupant behaviour. These scenarios consider both small, localized fires and large, fast-spreading fires that could overwhelm firefighting efforts.

Worst Case Scenarios

In addition to more likely fire events, engineers also model worst-case scenarios to assess how the building would perform under extreme conditions. For example, a worst-case scenario might assume that fire suppression systems fail, or that the fire starts in a high-risk area where it could spread quickly. The goal is to ensure that even in these scenarios, the building's design provides sufficient time for safe evacuation and minimizes structural damage.

Performance Assessment

Once scenarios are modelled, the performance of the building is evaluated under each condition. Engineers assess factors such as how quickly the fire spreads, how much time is available for occupants to evacuate, and how well the building's fire protection systems respond.

Fire and Smoke Dynamics

Critical Components

One of the critical components of performance-based fire safety design is understanding fire and smoke dynamics, how fire and smoke behave in a building during a fire. To predict this behaviour, engineers use advanced computational tools such as computational fluid dynamics (CFD) models.

CFD Modelling

CFD simulations model the movement of fire and smoke throughout the building. They consider factors such as the layout of the building, airflow. The location of doors and windows, ventilation systems, and the type of materials involved in the fire. CFD models can show how quickly smoke fills different areas of the building, which escape routes remain passable, and where heat or toxic gases might accumulate.

Impact on Occupants

Fire and smoke dynamics are particularly important when assessing occupant safety. Smoke inhalation is often more dangerous than the fire itself, so understanding how smoke will move and accumulate helps engineers design better smoke control systems. For example, pressurization systems in stairwells or smoke extraction systems in large atrium spaces may be modelled and optimized to ensure that occupants can safely evacuate.

Impact on Structural Integrity

Understanding fire and smoke behaviour also helps engineers assess the impact of the fire on the building's structure. Hot gases can weaken structural components, especially in steel-framed buildings, so engineers need to model how fire exposure will affect load-bearing walls, floors, and columns.

Evacuation Modelling

A central element of performance-based fire safety design is the evacuation strategy. To ensure that the building can be evacuated safely in the event of a fire, engineers use evacuation modelling tools to simulate how long it will take occupants to leave the building under different conditions.

Required Safe Egress Time (RSET)

This is the amount of time that it takes for all occupants to safely exit the building once a fire has been detected. RSET includes factors such as the time it takes for occupants to become aware of the fire, their reaction time, the speed at which they move through the building, and the number and location of exits.

Available Safe Egress Time (ASET)

ASET is the amount of time that occupants must safely evacuate before conditions become untenable (e.g., before smoke fills escape routes or temperatures rise to dangerous levels). ASET depends on the fire and smoke dynamics, the effectiveness of fire detection and suppression systems, and the building's layout.

Comparing RSET and ASET

Engineers compare RSET with ASET to ensure that there is enough time for safe evacuation. The design goal is to make sure that RSET is always less than ASET, meaning that occupants can evacuate safely before fire and smoke conditions become life-threatening. If RSET exceeds ASET, the design must be modified—either by improving evacuation routes, adding more exits, or enhancing smoke control and fire suppression systems.

Structural Fire Resistance

Is another critical component of performance-based fire safety design. This aspect focuses on ensuring that the building's structural elements, such as beams, columns, floors, and walls—can withstand fire long enough to allow for evacuation and firefighting efforts, and to prevent collapse.

Evaluating Structural Performance

Engineers use performance-based approaches to assess how different structural materials and systems will behave under fire conditions. For instance, steel loses strength at elevated temperatures, so fire protection measures like fireproof coatings or concrete encasement might be required. Concrete can crack or spall under intense heat, so designers may need to use special fire-resistant mixes or additional reinforcement.

Risk of Structural Collapse

In high-rise buildings, large open spaces, or industrial buildings like warehouses, the risk of structural collapse during a fire can be significant. Engineers model how the fire will affect critical load-bearing elements and identify any points of weakness. In some cases, performance-based designs may involve compartmentation, dividing the building into fire-resistant compartments to prevent the fire from spreading and reducing the load on structural elements.

Fire Resistance Ratings

Instead of using prescriptive fire resistance ratings for all parts of the building, performance-based design allows engineers to focus fire protection measures on the most critical areas. For example, key structural components might be designed to withstand fire for 60, 90, or 120 minutes, depending on the expected fire scenario and the time required for evacuation.

Fire Safety Performance-Based Design

Fire Scenarios

Engineers model a range of fire scenarios, from minor fires to worst-case events, to ensure that the building can handle a variety of conditions. This allows for a more tailored fire safety strategy.

Fire and Smoke Dynamics

Using tools like CFD, engineers predict how fire and smoke will behave in the building. This helps them design systems that control smoke, protect occupants, and prevent structural damage.

Evacuation Modelling

By simulating evacuation times and comparing them to the time available before conditions become dangerous, engineers can design safe, effective evacuation routes and strategies.

Structural Fire Resistance

Performance-based design evaluates how the building's structure will withstand fire, ensuring that it remains intact long enough for evacuation and firefighting efforts. This may involve tailored fire protection measures for critical components.

Through performance-based fire safety design, buildings can be equipped with more effective, flexible, and customized fire protection strategies that align with their unique characteristics and uses.

Chapter 12

Building Materials – Properties and Performance

Fire safety in buildings is a crucial consideration in design, construction, and material selection. Fire engineers must assess how materials behave under fire conditions to minimize risks to life, property, and structural integrity. This involves understanding combustibility, reaction to fire, fire resistance, and the effects of fire on different construction materials.

FIRE ENGINEERING PRINCIPLES FOR MATERIALS SELECTION

Fire engineering evaluates materials based on:

- **Reaction to Fire** – How materials contribute to fire spread.
- **Fire Resistance** – The ability to maintain structural integrity.
- **Smoke and Toxicity** – Emission of harmful gases.
- **Thermal Conductivity** – Heat transfer within materials.
- **Structural Performance** – Load-bearing capacity under fire conditions.

Common Building Materials and Their Fire Performance

Concrete

- **Fire Resistance**: High; does not burn or emit toxic fumes.
- **Thermal Conductivity**: Low; acts as a heat insulator.
- **Failure Mode**: Spalling at high temperatures (>600°C), reducing strength.
- **Improvement**: Fire-resistant coatings and fibres (e.g., polypropylene fibres) reduce spalling.

Steel

- **Fire Resistance**: Poor; loses strength above 500°C.
- **Thermal Conductivity**: High; heat spreads rapidly.
- **Failure Mode**: Structural failure due to softening.
- **Improvement**: Fireproof coatings (intumescent paints), encasement in concrete or gypsum.

Timber
- **Fire Resistance**: Limited; ignites at ~300°C.
- **Thermal Conductivity**: Low, burns predictably with char layer formation.
- **Failure Mode**: Loss of cross-section affects load bearing.
- **Improvement**: Fire-retardant treatments engineered wood (CLT with fire-rated adhesives).

Brick and Masonry
- **Fire Resistance: High; withstands high temperatures without combustion.**
- **Thermal Conductivity: Moderate; resists fire spread.**
- **Failure Mode**: Cracking and weakening under prolonged fire exposure.

Glass
- **Fire Resistance: Low; shatters when exposed to rapid temperature changes.**
- **Thermal Conductivity: High.**
- **Improvement**: Fire-rated glass (wired, laminated, intumescent).

Gypsum Plasterboard
- **Fire Resistance: High; contains water that slows fire spread.**
- **Failure Mode: Breaks down under prolonged fire exposure.**
- **Improvement**: Fire-rated gypsum boards with additives.

Plastics and Composites
- **Fire Resistance**: Poor; most are combustible.
- **Smoke & Toxicity**: Release toxic gases.
- **Improvement**: Fire-retardant additives and encapsulation.

Fire Retardant Treatments and Coatings
- **Intumescent Coatings**: Expand when heated, forming an insulating char layer.
- **Flame Retardants**: Added to plastics, wood, and textiles to reduce flammability.
- **Encasement Systems**: Using gypsum, cement, or boards to shield materials.

Fire Testing Standards and Classifications
- **BS 476 (UK Standard)** – Fire tests on building materials.
- **EN 13501-1 (European Standard)** – Reaction to fire classification (A1–F).
- **ASTM E119 (US Standard)** – Fire resistance of construction materials.

Material selection in fire engineering must balance fire resistance, structural performance, and regulatory compliance. Advances in fire-rated materials, coatings, and passive fire protection significantly improve building safety

FIRE RESISTANCE vs. REACTION TO FIRE

In fire engineering, fire resistance and reaction to fire are two key properties that determine how materials behave in a fire. While fire resistance measures a material's ability to maintain its function during a fire, reaction to fire assesses how a material ignites, contributes to fire spread, and releases smoke or toxic gases. These properties influence fire safety strategies, building codes, and material selection.

Fire Resistance: The Ability to Withstand Fire

Definition

Fire resistance refers to the ability of a material or building element (e.g., walls, floors, doors, columns, beams) to withstand fire without structural failure for a specified period, typically measured in minutes (e.g., 30, 60, 90, 120, 180 minutes).

Performance Criteria

Fire resistance is assessed based on three main factors:

- **Load-bearing capacity (R – Resistance):** The ability to maintain structural integrity under fire conditions.
- **Integrity (E – Enclosure):** The ability to prevent flames and hot gases from passing through.
- **Insulation (I – Insulation):** The ability to limit heat transfer to the unexposed side.

Materials and structures are given a fire resistance rating based on standardized fire tests, such as:

- **BS 476 (UK)**
- **EN 1363-1 (Europe)**
- **ASTM E119 (USA)**

Fire Resistance Ratings

- **30 minutes** – Delays fire penetration for half an hour.
- **60 minutes** – Provides structural integrity for one hour.
- **120 minutes** – Suitable for high-rise buildings and critical infrastructure.
- **180+ minutes** – Used in fire-safe compartments for extreme fire protection.

Examples of Fire-Resistant Materials	
Material	**Fire Resistance Features**
Concrete	High resistance; does not burn but may spall under heat.
Steel (with protection)	Requires coatings or encasement; loses strength above 500°C.
Brick/Masonry	Withstands high temperatures; does not combust.
Fire-Rated Glass	Remains intact for a set duration under fire exposure.
Gypsum Board	Contains water that helps delay fire spread

Methods to Improve Fire Resistance
- **Passive fire protection (PFP):** Fire-resistant coatings, intumescent paints.
- **Fireproof cladding and insulation:** Materials that reduce heat transfer.
- **Encasement of structural elements:** Concrete or fire-rated boards around steel beams.

Reaction to Fire: How Materials Behave in a Fire
Definition
Reaction to fire describes how a material behaves when exposed to fire, including:
- Ignition temperature (how easily it catches fire).
- Flame spread rate (how quickly fire spreads).
- Heat release rate (how much energy it contributes to the fire).
- Smoke and toxic gas production.

Reaction to Fire Testing
Reaction to fire is assessed using standards such as:
- **EN 13501-1 (European classification)**
- **BS 476 Part 6 & 7 (UK)**
- **ASTM E84 (Steiner Tunnel Test in the US)**

Classification of Materials (EN 13501-1)

Class	Reaction to Fire Behaviour
A1	Non-combustible (e.g., concrete, stone, brick).
A2	Very limited combustibility (e.g., fire-resistant glass).

Class	Reaction to Fire Behaviour
B	Low flammability (e.g., treated wood, fire-rated plastics).
C	Moderate combustibility (e.g., some timber-based panels).
D	Easily combustible (e.g., untreated wood, textiles).
E	Highly flammable (e.g., plastics, foam insulation).
F	No fire performance determined.

Smoke and Toxicity Ratings

Materials are also classified based on **smoke production (s1, s2, s3)** and **burning droplets (d0, d1, d2)**:

- **s1:** Low smoke production.
- **s2:** Moderate smoke production.
- **s3:** High smoke production.
- **d0:** No flaming droplets.
- **d1:** Some flaming droplets.
- **d2:** Many flaming droplets.

Examples of Material Reactions to Fire

Material	Reaction to Fire Characteristics
Timber (untreated)	Combustible, moderate flame spread; forms char layer.
Steel	Does not ignite but weakens under high heat.
Plastics (e.g., PVC, foam insulation)	Highly flammable, releases toxic smoke.
Gypsum Board	Low combustibility due to water content.
Glass	Does not ignite but may shatter.

Methods to Improve Reaction to Fire

- **Flame retardants:** Chemicals added to reduce ignition and flame spread.
- **Fire-resistant coatings:** Intumescent paints or fireproofing sprays.
- **Encapsulation:** Covering combustible materials with non-flammable layers.

Key Differences Between Fire Resistance and Reaction to Fire

Aspect	Fire Resistance	Reaction to Fire
Definition	Ability to maintain structure under fire conditions.	Behaviour when exposed to fire (ignition, spread, smoke).
Measurement	Rated in minutes (30, 60, 90, etc.).	Rated by combustibility (A1–F) and smoke/toxicity.
Key Properties	Load-bearing capacity, integrity, insulation.	Ignitability, flame spread, heat release, smoke emission.
Testing Standards	ASTM E119, BS 476, EN 1363-1.	ASTM E84, BS 476 Part 6 & 7, EN 13501-1.
Examples of High Performance	Concrete (high fire resistance).	A1-rated materials (non-combustible, low smoke).
Examples of Poor Performance	Unprotected steel (collapses under heat).	Plastics, untreated wood (high flammability).

Conclusion

- **Fire Resistance** is crucial for structural integrity in buildings, ensuring they remain stable during fire events.
- **Reaction to Fire** is important for fire spread prevention, reducing ignition risk and controlling smoke/toxicity.
- **A balanced approach** in material selection considers both fire resistance (structural stability) and reaction to fire (flammability, smoke, and toxicity) to enhance overall fire safety.

Concrete Structures Reinforcement Calculation Methods

The design and analysis of reinforced concrete structures require rigorous calculation methods to ensure structural integrity. Some of the key calculation methods include:

- **Limit State Design (LSD):** Ensures safety by considering ultimate and serviceability limit states.
- **Working Stress Method (WSM):** Based on elastic theory, where stresses are kept within permissible limits.
- **Ultimate Load Method (ULM):** Considers the maximum load a structure can withstand before failure.

Reinforcement calculations include:

Moment Capacity (φMn) Calculation:

$$Mn = A_s f_y (d - a/2)$$

where:

A_s is the area of steel reinforcement,

f_y is the yield strength of reinforcement,

d is the effective depth,

a is the depth of the equivalent rectangular stress block.

Shear Capacity Calculation: Utilizes the shear strength formula:

$$V_c = a\sqrt{(f_c')} bd$$

where:

f_c' is the concrete compressive strength,

b is the width of the section,

d is the effective depth,

a is a factor dependent on concrete properties.

Yield Strength of Reinforcement Bars

The yield strength of reinforcement bars (rebar) varies based on material properties and grades. Typical values include:

- **Mild Steel:** 250 MPa
- **High Yield Strength Deformed Bars (HYSD):** 415 MPa to 500 MPa
- **Thermo-Mechanically Treated Bars (TMT):** 500 MPa to 600 MPa

The selection of rebar depends on the structural application and required durability.

Stiffness of Steel

The stiffness of steel in reinforced concrete structures is primarily defined by Young's Modulus: $E_s = 200 Pa$. This high modulus ensures that steel reinforcement effectively carries tensile stresses while maintaining structural integrity.

Concrete Properties

Compressive Strength of Concrete

The compressive strength of concrete is a critical property influencing structural performance. Typical strengths include:

- **Normal Strength Concrete (NSC):** 20 MPa - 40 MPa
- **High Strength Concrete (HSC):** 50 MPa - 100 MPa
- **Ultra-High-Performance Concrete (UHPC):** > 150 MPa

The compressive strength is determined by standard cube or cylinder tests conducted at 28 days.

Stiffness of Concrete

Concrete stiffness is represented by its **modulus of elasticity** (E_c): $E_c = 4700\sqrt{(F_c')}$

Where:

F_c' is the characteristic compressive strength. For normal concrete, E_c ranges from 20 GPa to 40 GPa.

Concrete Structures

Beams in Bending

Concrete beams resist bending through the interaction of tensile reinforcement and concrete compression. The bending moment capacity is given by:

$$M_n = A_s f_y (d - a/2)$$

where:

- **Compression zone** in concrete resists compression forces,
- **Tensile reinforcement** carries the tensile stresses.

Failure modes include:

- Flexural failure (yielding of steel),
- Compression failure in concrete.

Shear Capacity in Beams

Shear failure in beams occurs due to insufficient shear strength in concrete or lack of stirrups. The shear capacity is calculated using:

$$V_n = V_c + V_s$$

where:

- **Concrete shear contribution:** $V_c = 0.17\sqrt{(f_c')}bd$
- **Stirrups shear contribution:** $V_s = A_v f_y d / s$, where A_v is stirrup area and s is spacing.

Unreinforced Concrete Column

Unreinforced concrete columns rely entirely on concrete for load bearing and are suitable for low-load applications. The axial strength is calculated as: $P_n = 0.85 f_c' A_g$ where A_g is the gross cross-sectional area.

Reinforced Concrete Column

Reinforced concrete columns are categorized into:

- **Short Columns:** Primarily fail due to crushing.
- **Slender Columns:** Prone to buckling.

The load-carrying capacity is: $P_n = 0.85 f_c' \left(A_g - A_s \right) + f_y A_s$ where A_s is the reinforcement area.

Concrete structures rely on reinforcement for tensile strength and stability. Key parameters such as yield strength, stiffness, and compressive strength play a critical role in structural performance. Understanding bending, shear, and column behaviour is essential for safe and efficient design in reinforced concrete structures.

Steel Structures

Properties of Steel in Fire

Steel is widely used in structural applications due to its high strength, ductility, and ease of fabrication. However, in fire conditions, its mechanical properties degrade significantly, impacting structural stability.

Mechanical Properties of Steel at Elevated Temperatures

The mechanical properties of steel deteriorate with increasing temperature. According to Eurocode 3 (EN 1993-1-2), the main properties affected are:

- **Yield Strength (σy)**
- **Elastic Modulus (E)**
- **Thermal Expansion**

Stress-Strain Relationship at Elevated Temperatures

The reduction factors for yield strength and modulus of elasticity with temperature are:

Temperature (°C)	Yield Strength Reduction Factor (k_y)	Elastic Modulus Reduction Factor (k_E)
20	1.00	1.00
200	0.90	0.95
400	0.70	0.80
600	0.50	0.40
800	0.10	0.15
1000	0.00	0.00

Steel begins to lose its load-bearing capacity at around 500-600°C with collapse at around 800°C, making fire protection critical.

Steel Heating: Surface Heat Transfer and Emissivity

When exposed to fire, steel temperature rises due to radiative and convective heat transfer from the surrounding fire gases.

Surface Heat Transfer Coefficient (αc)

Heat transfer occurs through:

Radiation (αr): Exchange of heat between fire and steel surface.

Convection (αc): Heat transfer from hot gases to steel surface.

For standard fires, Eurocode 3 (EN 1993-1-2) provides values:

Convection coefficient: $a_c = 25 w/m^2 k$

Emissivity of steel: $\varepsilon_m = 0.7$ for oxidized steel

Section Factor (A/V)

The section factor (A/V) is the ratio of the surface area exposed to fire (A) to the volume (V) of the steel cross-section. It influences the heating rate of steel.

$$\frac{A}{V} = \frac{Exposed Perimeter(m)}{Cross\,Sectional Area(m^2)}$$

Higher (A/V) ratios result in faster heating, making slender sections more vulnerable.

Heating Curve of Steel (Uninsulated and Insulated)

Uninsulated Steel Temperature Development

The temperature of unprotected steel exposed to ISO 834 standard fire can be calculated using:

$$\frac{dT_s}{dt} = \frac{q}{p_s c_s V}$$

Where:

T_s = steel temperature (°C)
P_s = density of steel $(7850 kg/m^3)$
C_s = specific heat capacity $(600 j/kgK)$
V = volume of steel section
q = heat flux (Wm^2)

$$T_s(t) = T_o + \frac{1}{\rho_s c_s} \int_0^t q \, dt$$

Insulated Steel

Insulated steel heats up much slower, as the insulation reduces heat transfer. The thermal resistance of fire protection material is given by:

$$R = \frac{d}{\lambda}$$

Where:

d = insulation thickness (m)

λ = thermal conductivity (W/mK)

The fire resistance period of insulated steel is determined based on required critical temperature, usually 550°C.

Verification of Steel Beams and Columns (Eurocode 3, EN 1993-1-2)

Verification of Steel Beams

The moment resistance of a steel beam at elevated temperature is given by:

$$M_{fi,Rd} = k_y(\emptyset).M_{pl,Rd}$$

Where:

$k_y(\emptyset)$ = reduction factor for yield strength at temperature $(\emptyset)$

$M_{pi,Rd}$ = plastic moment resistance at room temperature

$$M_{pl,Rd} = \frac{f_y.W_{pl}}{\gamma m}$$

Example Calculation: For an **IPE 300** section $\left(w\_pl = 4.38 \times 10^{-3} m^3\right)$ at 600°C:

$$M_{pl,Rd} = \frac{10.50.355 \times 10^6 \times 4.38 \times 10^{-3}}{1}$$

$$M_{fi,Rd} = 0.50 \times M_{pl,Rd}$$

Verification of Steel Columns

Columns are verified for axial compression resistance:

$$N_{fi,Rd} = k_y(\emptyset).x.\frac{A.f_y}{\gamma m}$$

Where:

x = buckling reduction factor (from buckling curve)

A = cross-sectional area

For **HEB 300** $\left(A = 11.54 \times 10^3 \, m^2\right)$ at **600°C**:

$$N_{fi,Rd} = 0.50 \times 0.8 \times \frac{11.54 \times 10^{-3} \times 355 \times 10^6}{1.0}$$

Charts and Calculations

The following section contains heating curves for uninsulated and insulated steel and includes load-bearing capacity reduction charts.

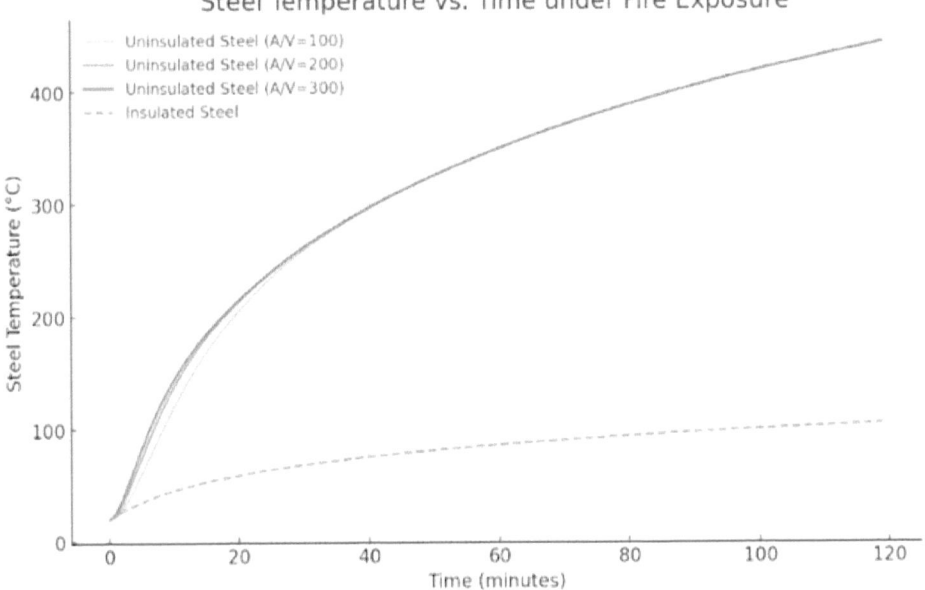

The chart above illustrates the heating curves of steel under fire conditions:

Uninsulated steel with different section factors (A/V):

Higher A/V ratios lead to **faster temperature rise**.

By **30 minutes**, unprotected steel can reach **600°C**, severely reducing its strength.

Insulated steel

Temperature rise is significantly slower, staying below 300°C even after 120 minutes.

The following is a load-bearing capacity reduction chart for steel beams and columns at elevated temperatures.

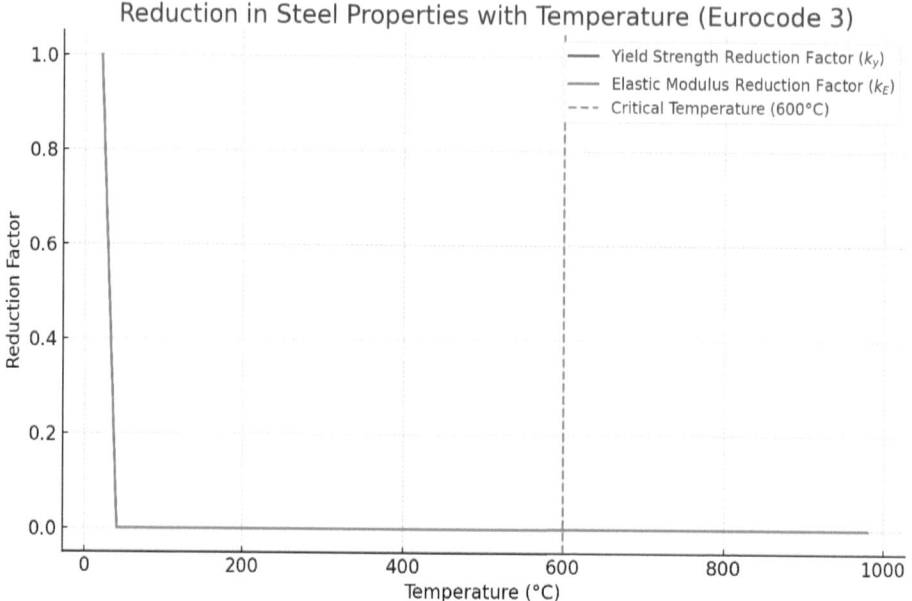

The chart above illustrates the reduction factors for yield strength k_y and elastic modulus (k_e) of steel as temperature increases:

Yield strength (k_y):
- Drops to **50% at 600°C**.
- At **800°C**, it falls to **10%**, making the structure highly vulnerable.

Elastic modulus (k_e):
- Reduces to **80% at 400°C** and just **40% at 600°C**.
- At **800°C**, only **15%** of stiffness remains.

Critical threshold (600°C):
- Many design codes consider **550-600°C** as the failure temperature for unprotected steel.

Conclusion

Unprotected steel can lose structural integrity in under 30 minutes in a fire.

Fire insulation significantly delays temperature rise, improving safety.

Eurocode 3 provides reduction factors to verify fire resistance of beams and columns.

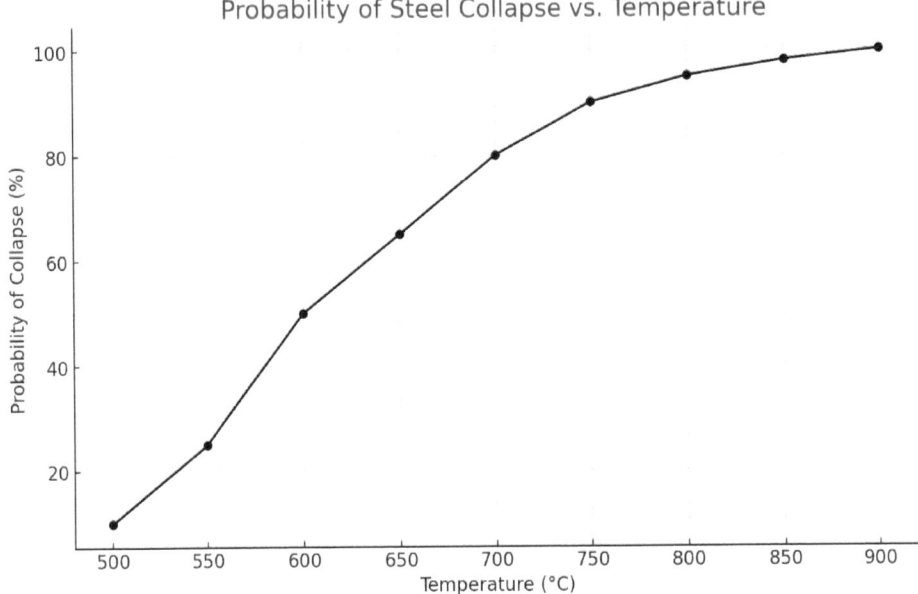

The chart above shows the fire resistance time estimates based on insulation thickness and the probability of steel collapse due to temperature.

It can be seen that while the steel loses its integrity at around 800°C total collapse results between 850°C and 900°C

Wooden Structures

Wood is a combustible material widely used in construction due to its availability, sustainability, and structural properties. However, its behaviour in fire conditions requires thorough analysis to ensure safety. Fire engineering design methods focus on calculating charring rates, strength reduction, and verifying load-bearing capacity in fire scenarios.

Calculation Methods

Charring Layer

When exposed to fire, wood undergoes pyrolysis, forming a char layer that insulates the remaining cross-section. The rate of charring depends on the fire exposure, species of wood, and environmental conditions.

Reduced Cross-Section Method

This method calculates the residual load-bearing capacity of a fire-exposed wooden element by reducing its cross-section based on the expected charring depth over time.

Notional Charring Rates

The charring rate (β_n) represents the depth at which wood is converted to char per unit of time. Common values under standard fire conditions are:

- Softwood: 0.65 mm/min
- Hardwood: 0.5 mm/min
- Laminated timber: 0.7 mm/min

The depth of the char layer (d_{char}) can be calculated as: $d_{char} = \beta_n \times t_{fire}$ Where t_{fire} is the fire duration in minutes.

Pyrolysis Zone

The pyrolysis zone is the transitional layer between the charred and unburnt wood, where thermal degradation and strength reduction occur. This zone plays a role in structural calculations.

Strength Reduction Method

The residual strength of timber elements is affected by fire exposure. Monodimentional charring rates help in determining the remaining strength by accounting for the depth of char formation.

The effective cross-section (A_{eff}) is calculated as: $A_{eff} = A_{orig} - d_{char} \times b$ is the original cross-section and A_{orig} is the breadth of the element.

Rounded Corners

For elements with rounded corners, fire exposure accelerates charring in these areas due to increased heat transfer, requiring additional considerations in design calculations. The increased charring depth at corners can be estimated using a correction factor k_c, where: $d_{char,corner} = k_c \times d_{char}$ Typical values of k_c range from 1.2 to 1.5 depending on the corner radius.

Strength and Stiffness Reduction Factor

Fire exposure reduces both strength and stiffness of wood. The reduction factors depend on the temperature profile, fire duration, and timber species. The strength reduction factor (k_{fire}) is obtained from: $k_{fire} = \dfrac{f_{m,fire}}{f_m}$ where $f_{m,fire}$ is the fire-modified strength and f_m is the strength at ambient conditions.

Parametric Fire Design of Wood Structures

Maximum Duration of Burning

Parametric fire design considers the duration of burning based on ventilation, compartment size, and fire load density.

Charring Rate in Parametric Fires

In real fire scenarios, the charring rate varies with time and fire severity. The charring rate increases in the growth phase and stabilises in the fully developed fire phase.

Charring Layer in Parametric Fires

The thickness of the char layer in parametric fires is calculated using time-dependent charring rates, considering fire intensity fluctuations. For example, if a fire follows a parametric temperature-time curve with a peak temperature of 1000°C, the charring rate can be adjusted dynamically: $d_{char, parametric} = \int_0^{t_{fire}} \beta_n(t) dt$ where $\beta_n(t)$ is the time-dependent charring rate.

Verification of Wooden Beams

Wooden Beams in Fire

Beams are assessed for fire resistance based on reduced cross-section and residual strength. The moment resistance is calculated as: $M_{Rd} = W_{eff} \times f_{m,fire}$ where W_{eff} is the effective section modulus.

Laminated Wood Beams

Laminated timber beams exhibit different charring rates due to glue-line degradation. Additional protection may be required, such as:

- Protective cladding with fire-rated gypsum board
- Fire-resistant coatings or intumescent paints
- Larger cross-sections to allow sacrificial charring layers

Verification of Wooden Columns

Slenderness Ratio

The slenderness ratio (λ) is the ratio of column length to its effective cross-sectional radius of gyration: $\lambda = \dfrac{L}{i}$ where i is the radius of gyration.

Euler Stress

Euler's critical buckling stress (σ_{cr}) is calculated using: $\sigma_{cr} = \dfrac{\pi^2 E}{\left(L_{eff}/i\right)^2}$ where

E is the modulus of elasticity and L_{eff} is the effective column length.

Euler Factor

The Euler factor accounts for material and geometrical imperfections affecting column stability.

Relative Slenderness Ratio

The relative slenderness ratio (λ_{rel}) is calculated using: $\lambda_{rel} = \dfrac{\lambda}{\lambda_{cr}}$ where λ_{cr} is the critical slenderness ratio.

The k_{fire} coefficient adjusts mechanical properties under fire exposure, based on experimental data.

Critical Buckling Factor

The critical buckling factor (x_{fire}) accounts for stability reduction in fire conditions, calculated as $x_{fire} = \dfrac{1}{1+\lambda_{rel}^{2.5}}$

Characteristic Resistance of Wood Columns

Characteristic resistance is determined by $F_{Rd} = \dfrac{A_{eff} f_{m,fire}}{\gamma_{m,fire}}$ where A_{eff} is the effective cross-section, $f_{m,fire}$ is the fire-modified strength, and $\gamma_{m,fire}$ is the safety factor.

Fire engineering design of wood structures ensures their integrity during fire exposure by incorporating charring rates, residual strength calculations, and stability assessments. Proper design and verification methods help enhance fire resistance and ensure structural safety.

References

Eurocode 5: Design of Timber Structures – Part 1-2: General – Structural Fire Design

Fire safety regulations: Building Regulations Approved Document B (UK)

Scientific papers

"Fire Resistance of Timber Structures" – Structural Engineering International

"Charring Rates of Different Wood Types Under Parametric Fire Conditions" – Fire Safety Journal

"Performance of Laminated Timber in Fire" – Journal of Structural Fire Engineering

Fire Loads on Structures

Fire load is a critical factor in fire safety engineering and structural fire protection. It represents the total potential energy released by the combustion of combustible materials in a given space, affecting fire severity, spread, and structural integrity. Understanding fire loads is essential for designing fire-resistant buildings and ensuring safety compliance.

Definition of Fire Load

Fire load is typically defined as:

$$\text{Fire Load (Mj)} = \frac{\Sigma \text{ mass of combustible materials} \times \text{heat of combustion}}{\text{Floor area (m}^2\text{)}}$$

Alternatively, it can be expressed in terms of weight:

$$\text{Fire Load (kg/m}^2\text{)} = \frac{\text{mass of combustibles (kg)} \times \text{calorific value}}{\text{Floor area (m}^2\text{)}}$$

The fire load density (FLD) is often expressed in megajoules per square meter (MJ/m²), representing the potential energy available per unit floor area.

Types of Fire Loads

Fire loads can be categorized into different types based on the source and nature of combustibles:

Fixed Fire Load

These are part of the building structure or permanently attached materials.

Examples include:
- Timber elements (roofs, floors, and walls in timber-framed structures).
- Insulation materials.
- Gypsum boards (though not directly combustible, they influence fire behaviour).
- Fire-retardant coatings.

Movable Fire Load

These are materials that can be introduced or removed from a space.

Examples:
- Furniture (wooden, plastic, or upholstered).
- Paper, books, and textiles.
- Storage materials (pallets, packaging, fuels).

Operational Fire Load

Includes fuel sources linked to the functioning of equipment.

Examples:
- Petrol or oil in machinery.
- Chemical storage in laboratories.
- Flammable liquids in industrial settings.

Factors Affecting Fire Load

Several parameters influence the fire load in a structure:

Type of Materials

Different materials have varying calorific values.
- Wood: ~17-20 MJ/kg
- Plastic: ~30-40 MJ/kg
- Paper: ~15-20 MJ/kg
- Gasoline: ~44 MJ/kg

Quantity of Combustibles

Larger amounts of combustibles increase the fire load, leading to higher energy release and prolonged burning.

Distribution and Arrangement
- The spatial distribution of combustibles affects fire spread.
- Clusters of high-density fire loads may lead to flashover conditions.

Ventilation
- Oxygen availability influences combustion efficiency.
- High ventilation can lead to increased burning rates, while limited oxygen can lead to smouldering fires and toxic smoke accumulation.

Fire Protection Measures

Fire-resistant materials, suppression systems, and compartmentalization reduce fire spread and mitigate fire load impact.

Fire Load Classification

Fire loads can be classified based on their intensity:

Fire Load Density (MJ/m²)	Fire Load Classification	Examples
< 200 MJ/m²	Low Fire Load	Office spaces, residential homes.
200 – 400 MJ/m²	Moderate Fire Load	Hotels, schools, shopping malls.
400 – 800 MJ/m²	High Fire Load	Industrial warehouses, factories.
> 800 MJ/m²	Very High Fire Load	Petrochemical plants, chemical storage.

Impact of Fire Load on Structural Performance

The fire load directly influences the severity and duration of a fire, affecting structural integrity in various ways:

Thermal Effects on Materials

Steel: Weakens at ~500°C and loses ~50% of its strength at 600°C.

Concrete: Spalling may occur due to moisture expansion.

Timber: Burns at ~0.6 mm/min charring rate, reducing cross-sectional integrity.

Fire Spread and Flashover

- Higher fire loads increase the likelihood of flashover, where all combustibles ignite simultaneously.
- Structures with high fire loads require advanced compartmentalization and active suppression strategies.

Load-Bearing Capacity Reduction

Structural elements exposed to fire experience **thermal expansion**, leading to:

- Buckling in steel.
- Cracking in concrete.
- Charring in timber, reducing its load-bearing capacity.

Smoke and Toxic Gas Emission

High fire loads contribute to greater smoke production, reducing visibility and increasing toxicity risks (e.g., carbon monoxide, hydrogen cyanide).

Fire Load Management Strategies

To mitigate fire hazards, several strategies can be implemented:

Fire-Resistant Materials

- Use of fire-resistant coatings, intumescent paints, and non-combustible insulation.

Compartmentalization
- Fire-rated walls and doors limit fire spread.
- Use of fire zones in large industrial or commercial buildings.

Active Fire Protection
- Automatic sprinklers, fire suppression systems, and extinguishers reduce fire intensity.
- Smoke control and extraction systems improve visibility for evacuation.

Storage and Housekeeping
- Limiting combustible storage in key areas.
- Keeping escape routes free from fire loads.

Fire Load Calculation and Assessment

Regular fire risk assessments based on fire load surveys.

Implementation of building codes and regulations (e.g., NFPA 557, Eurocode 1).

Fire Load Standards and Regulations

Fire load management is guided by international standards, including:

- **NFPA 557** – Standard for determining fire loads for buildings.
- **Eurocode 1 (EN 1991-1-2)** – Fire load density assessment in structures.
- **BS 9999** – Fire safety in the design, management, and use of buildings.
- **ISO 834** – Fire resistance tests for structural elements.

Fire loads play a vital role in fire safety engineering, determining fire severity, spread, and structural vulnerability. Proper assessment and management of fire loads are essential to ensuring building resilience, occupant safety, and regulatory compliance. By incorporating fire-resistant materials, compartmentalization, and suppression systems, structures can withstand fire hazards more effectively.

Chapter 13

Building Compartmentation

Fire compartmentation is a fundamental fire protection strategy in building design that involves dividing a structure into separate compartments to contain the spread of fire, smoke, and heat. This passive fire protection method is crucial for life safety, property protection, and maintaining structural stability.

From a fire engineering perspective, compartmentation integrates fire-resistant walls, floors, ceilings, and doors to control fire growth, facilitate safe evacuation, and limit damage to critical building infrastructure.

It also plays a role in ensuring compliance with national and international fire codes and standards, such as the Building Regulations (UK Approved Document B), NFPA 101 (USA), and Eurocode EN 1991-1-2.

Fire engineering judgment is crucial when determining compartment sizes in buildings with increased fire loads because standard prescriptive fire safety codes may not adequately address unique risks associated with high fire loads.

Some of the key reasons why fire engineers must apply their expertise in such situations:

- **Fire Load Variability and Its Impact on Fire Behaviour**

  Fire load refers to the amount of combustible material within a given space, typically expressed in MJ/m^2. In some occupancies, such as warehouses, libraries, or data centres, the fire load can be significantly higher than in standard residential or office buildings.

- **Higher Fire Loads Lead to Longer Burning Times**: Increased fuel availability extends fire duration, requiring enhanced compartmentation strategies.

- **Greater Heat Release Rates (HRR)**: More combustible material means more intense fires, which can challenge standard compartment designs.

- **Possible Flashover and Early Structural Collapse**: If fire compartments are too large, the temperature can rise quickly, leading to rapid fire spread and potential structural failure.

- **Engineering Judgment Needed**
  Fire engineers should assess whether standard compartment sizes can contain the fire load or if modifications (such as enhanced fire-resistant materials or additional fire suppression systems) are necessary.

Objectives of Fire Compartmentation

The primary objectives of fire compartmentation in buildings are:
- **Life Safety:** Prevent fire and smoke from rapidly spreading, allowing occupants sufficient time to escape safely.
- **Property Protection:** Minimize structural damage and reduce fire impact on high-value areas.
- **Business Continuity:** Protect critical assets and services, ensuring a building remains operational after a fire incident.
- **Firefighting Operations:** Provide containment zones that allow firefighters to control and extinguish the fire effectively.
- **Compliance with Regulations:** Ensure adherence to fire safety laws and reduce liability risks.

Fire Engineering Principles Behind Compartmentation

Fire Growth and Spread

Fire spreads through conduction, convection, and radiation, and compartmentation aims to break this chain by restricting fire and smoke within designated zones. The key parameters include:
- **Flashover Prevention:** By limiting the oxygen and fuel available, compartmentation can delay or prevent the occurrence of flashover, where fire engulfs an entire compartment.
- **Heat Transfer Control:** Use of fire-resistant materials to reduce heat transmission across walls, floors, and ceilings.
- **Smoke Control Measures:** Design to prevent smoke leakage through door gaps, ventilation ducts, and service penetrations.

Fire Resistance Ratings

Materials used in compartmentation must meet fire resistance ratings expressed in minutes (e.g., 30/60/90/120 minutes) according to tests like BS EN 1364, ASTM E119, or ISO 834. These ratings define the time a material can withstand fire exposure while maintaining:

- **Structural Integrity** (ability to maintain load-bearing capacity)
- **Insulation** (preventing excessive temperature rise on the unexposed side)
- **Integrity** (preventing passage of flames and hot gases)

Compartment Size and Design

Building regulations define maximum compartment sizes based on occupancy type and fire load. For instance:

- **Residential Buildings:** Compartment sizes are typically smaller to limit fire spread and ensure safe escape routes.
- **Industrial Buildings:** Larger compartments may be permitted with active suppression (e.g., sprinklers).
- **Hospitals and Care Homes:** Require enhanced compartmentation for phased evacuation of vulnerable occupants.

Components of Compartmentation

Fire-Resisting Walls and Floors

- **Fire-resistant walls** (e.g., concrete, gypsum board with intumescent layers) provide barriers between compartments.
- **Fire-resistant floors/ceilings** ensure vertical containment.
- Fire Doors and Fire Shutters
- **Fire-rated doors** (e.g., FD30, FD60) prevent fire and smoke from spreading while allowing safe egress.
- **Self-closing mechanisms and intumescent seals** help maintain compartment integrity.
- **Service Penetrations and Fire Stopping**
- **Ducts, pipes, and cables** create vulnerabilities in compartment walls.
- **Fire dampers and intumescent seals** ensure airtight seals to prevent fire spread through service openings.

Smoke Barriers and Curtains

Used in atriums, shopping malls, and high-rise buildings to direct smoke away from escape routes.

Compartmentation Example: A 60m Residential High-Rise Building

Building Description

- **Height:** 60m (approximately 20 stories)

- **Use:** Residential (apartments/flats)
- **Occupant Load:** Approx. 500 residents
- **Construction Type:** Reinforced concrete structure with fire-resistant partitions
- **Fire Safety Strategy:** Passive and active fire protection integration

Fire Compartmentation Strategy

Apartment Compartmentation

- Each apartment must be a separate fire compartment with minimum 60 minutes fire resistance.
- Party walls and floors between apartments must have a minimum of 60 minutes fire resistance.

Stairwells and Escape Routes

- Fire-rated stairwells must be enclosed in 120-minute fire-resistant walls.
- Doors to stairwells must be FD60 self-closing fire doors with smoke seals.

Lifts and Service Shafts

- Lift shafts must have 120-minute fire resistance and be enclosed.

Corridors and Common Areas

- Corridors must be divided into compartments no longer than 30m with FD60-rated fire doors.

Basement & Parking Areas

- Basement car parks must be separated from residential areas by 120-minute fire-rated construction.

Roof and Plant Rooms

- Must be in separate 120-minute fire-rated enclosures.

External Wall Considerations

- **Non-combustible cladding and insulation materials** to prevent rapid fire spread.

Fire compartmentation remains a critical passive fire protection strategy in modern buildings. To ensure effectiveness:

- **Strict adherence to fire resistance ratings and building codes.**
- **Quality control in construction to prevent gaps and weak points.**
- **Regular fire risk assessments and maintenance of compartment elements.**
- **Integration with active fire protection for comprehensive safety.**

By combining engineering principles, regulatory compliance, and practical application, fire engineers can design buildings that protect lives and property effectively through robust fire compartmentation.

Chapter 14

Fire Investigation

Fire investigation is a specialized area of fire safety that focuses on determining the cause, origin, and development of a fire. It plays a critical role in understanding how fires start, whether any laws or regulations were violated, and in some cases, whether criminal activity was involved. The role and types of fire investigators, techniques for determining the fire cause, and the legal implications and reporting are described below.

Role of Fire Investigators

Fire investigators are highly trained professionals who examine the aftermath of a fire to establish how it began and spread. They work in a variety of contexts, including:

Post-fire scene examination

Fire investigators are typically called to a fire scene after the fire has been extinguished to investigate its cause. They collaborate closely with fire and rescue services, police, insurance companies, and forensic experts.

Determining the origin and cause

Their primary role is to identify the exact point of origin of the fire and determine what caused it. This could be an accidental ignition (e.g., electrical fault, careless use of fire), natural causes (e.g., lightning), or deliberate (arson).

Providing expert testimony

In cases of arson or fire-related criminal charges, fire investigators may be required to provide expert testimony in court. Their findings are critical to legal proceedings, helping to determine if there was foul play or negligence.

Assisting in the design of fire prevention measures

The findings from fire investigations can inform building design and fire safety regulations, helping to prevent similar incidents in the future. Fire investigators may work with fire engineers and safety officers to develop better fire prevention strategies based on the causes they identify.

Types of Fire Investigation Roles

Fire and Rescue Service Fire Investigators

Typically employed by fire services, these investigators focus on understanding how and why fires occur, with the goal of improving fire safety.

Police and Forensic Fire Investigators

Work on criminal cases involving arson or fire-related deaths, often in collaboration with forensic specialists and fire & Rescue service fire investigators.

Insurance Investigators

Employed by insurance companies to assess fire damage and determine whether claims are valid, especially in cases where fraud is suspected.

Private Fire Investigators

Hired by legal teams, corporations, or individuals to investigate specific fire incidents, often providing independent assessments in litigation or insurance cases.

Techniques For Determining Fire Cause

Fire investigators employ a range of techniques and scientific methods to determine the cause of a fire. These techniques are based on understanding fire behaviour, the materials involved, and the patterns left behind at the scene.

Fire investigators use multiple techniques to determine fire causes, including:

- **Fire Pattern Analysis**: Evaluating charring, smoke, and heat damage to trace the fire's point of origin.
- **Arc Mapping and Electrical Failure Analysis**: Examining electrical components for short circuits or arc faults.
- **Accelerant Detection and Chemical Analysis**: Using laboratory tests to identify traces of accelerants.
- **Heat and Flame Vector Analysis**: Studying burn damage intensity to determine fire movement.
- **Computational Fire Modelling**: Using software to simulate fire growth and validate findings.

Scene Examination

Fire scene examination begins with a systematic approach to identifying the fire's point of origin and cause:

Systematic Scene Analysis

The scene is documented with photographs and sketches, and investigators make observations about fire damage patterns, such as burn marks, soot deposits, and charring. This helps narrow down the point of origin.

Fire Patterns

Fire investigators look for patterns in the damage left by the fire, including:

- **V-patterns**: These are common patterns where the fire burns upward, leaving a V-shaped burn mark on vertical surfaces, with the apex of the V often pointing toward the point of origin.
- **Burn Intensity**: Areas with more intense burning can provide clues about the fire's point of origin, as the most intense heat is often near the fire's start.

Identifying Ignition Sources

Investigators look for potential ignition sources (e.g., electrical wiring, appliances, cigarettes, or heating equipment) near the fire's point of origin.

Fire Debris Analysis

Collection of Evidence

Fire investigators collect debris samples from the fire scene, particularly in areas where they suspect accelerants (such as petrol or other flammable liquids) may have been used in cases of arson.

Laboratory Analysis

These samples are sent to forensic labs for analysis. Techniques such as gas chromatography can detect trace amounts of accelerants, confirming their presence and identifying the exact substances used.

Electrical Fire Investigation

Electrical fires are a common cause of accidental fires. Fire investigators trained in electrical systems will:

Examine Wiring and Electrical Appliances

Investigators inspect the electrical systems and appliances at the fire scene, looking for signs of short circuits, overheating, or faulty components that could have sparked the fire.

Arc Mapping

Investigators may use arc mapping to trace the progression of electrical arcing in wiring systems. Arc faults can indicate whether the fire started due to an electrical malfunction or if the fire damaged the electrical system later.

Fire Dynamics and Modelling

Advanced fire investigation techniques may include the use of fire dynamics and fire modelling to recreate the fire scenario:

Fire Dynamics

Investigators analyse how fire and smoke behaved during the incident, factoring in ventilation, fuel load, and building materials. This helps them understand how the fire spread and whether any accelerants or unusual fire behaviours were involved.

Fire Modelling

Using software like Computational Fluid Dynamics (CFD) or fire simulation programs, investigators can create models of how the fire started and spread. These models help confirm or rule out different hypotheses about the fire's cause.

Witness Interviews : Gathering Eyewitness Accounts

Witnesses to the fire, such as building occupants or bystanders, are interviewed to gather information about what they saw or heard before, during, and after the fire. This can provide vital clues about potential causes, such as unusual smells (e.g., gasoline), sounds (e.g., electrical arcing), or suspicious behaviour (e.g., signs of arson).

Fire Scene Reconstruction

In some cases, investigators may reconstruct elements of the fire scene or ignition sources in controlled environments to understand how the fire may have started. This involves recreating conditions at the scene to test different hypotheses.

Fire scene reconstruction is a crucial aspect of fire investigation that involves piecing together physical and forensic evidence to determine how a fire started, spread, and ultimately was extinguished. The key components of fire scene reconstruction include:

- **Physical Evidence Correlation**: Comparing burn patterns, structural damage, and residual debris to establish fire spread direction and intensity.
- **Fire Growth Modelling**: Utilizing computational models and simulations to recreate the fire's development over time.
- **Temperature and Heat Analysis**: Examining the heat effects on materials to infer fire temperatures and durations at different points in the scene.
- **Witness and First Responder Reports**: Integrating accounts from occupants and emergency personnel to align the physical evidence with observed fire behaviour.
- **Environmental Influences**: Considering ventilation patterns, wind conditions, and material combustibility in reconstructing the fire.

- **Testing and Validation**: Conducting small-scale fire tests in controlled environments to compare real-world scenarios with the reconstructed fire model.

Example of Fire Scene Reconstruction

A warehouse fire investigation may begin with assessing charred remains and structural collapses. Investigators might identify the heaviest damage in the electrical control room, suggesting a possible electrical origin. Burn patterns on the floor and walls could indicate rapid fire spread due to flammable materials stored nearby. Using fire modelling software, investigators simulate the fire's progression, confirming that the ignition source was likely an overloaded circuit in an air conditioning unit. Witness statements from security personnel, coupled with forensic tests detecting arc mapping in electrical wiring, support the hypothesis. The final reconstruction report enables legal proceedings and insurance claim resolutions by providing a clear, scientifically backed explanation of the fire's cause and development.

Legal Implications and Reporting

Fire investigations often have significant legal implications, especially in cases where criminal activity or negligence is suspected. The investigator's findings can lead to legal action, influence insurance claims, and shape fire safety policies.

Legal Implications

Criminal Investigations and Arson

If a fire is suspected to be the result of deliberate ignition, it becomes a criminal matter. Fire investigators work closely with law enforcement to gather evidence that can be used in court. Arson is a serious offence in the UK, with penalties ranging from fines to life imprisonment, depending on the severity of the fire and whether it resulted in fatalities.

Negligence and Civil Liability

In cases where a fire resulted from negligence, such as poor maintenance of fire safety systems, faulty electrical wiring, or failure to comply with building regulations, fire investigators' reports can be used in civil cases. Responsible parties can face lawsuits for damages, injuries, or deaths caused by the fire.

Insurance Claims

Fire investigations are often crucial in determining the validity of insurance claims. If a fire is found to be the result of negligence or deliberate arson by the policyholder, the insurance company may deny the claim. Fire investigators working on behalf of insurers will focus on identifying any signs of foul play.

Fire Investigation Reports

A fire investigation report is a comprehensive document detailing the findings of the investigation. It includes:

- **Scene Description**: A description of the fire scene, including the extent of damage, the layout of the building, and any fire protection systems in place (e.g., sprinklers, alarms).
- **Point of Origin and Cause**: The exact location of the fire's origin and the suspected cause (e.g., electrical fault, accelerants, cooking appliances).
- **Fire Behaviour**: An analysis of how the fire developed and spread throughout the building, including any unusual fire behaviour or indications of accelerants.
- **Findings and Conclusions**: A summary of the key findings, including the cause of the fire, contributing factors, and any suspicious activity. The report will also note any violations of fire safety regulations or evidence of criminal activity.

Legal Standards and Best Practices

Fire investigators must adhere to strict legal standards and industry best practices when conducting investigations. These include:

- **Adherence to Protocols**: Following established protocols for evidence collection, handling, and storage to ensure the integrity of the investigation.
- **Expert Witness Testimony**: In legal cases, fire investigators may provide expert testimony, explaining their findings in court. Their role is to provide impartial, evidence-based conclusions.
- **Chain of Custody**: All evidence collected at the fire scene must be documented and preserved with a clear chain of custody to ensure it can be used in legal proceedings.

Fire investigation is a critical process for determining the cause of a fire, whether accidental, natural, or intentional. Fire investigators use a combination of scene examination, forensic techniques, and fire dynamics modelling to establish the origin and cause of a fire. Their work has profound legal implications, influencing criminal cases, civil liability, and insurance claims. Fire investigators play an essential role in the ongoing improvement of fire safety regulations, providing vital insights into how fires start and how future incidents can be prevented.

Chapter 15

Emerging Technologies in Fire Engineering

Emerging technologies in fire engineering are transforming how we approach fire safety, prevention, and response. The integration of innovative technologies enhances the effectiveness of fire safety measures and improves overall building safety. Below is a detailed account of several key areas in this field, including smart building technologies, fire modelling software and simulations, and future trends in fire safety solutions.

Smart Building Technologies

Smart building technologies leverage the Internet of Things (IoT), artificial intelligence (AI), and advanced sensor systems to enhance building safety, efficiency, and management.

Key Components:

IoT Sensors

These sensors monitor environmental conditions (e.g., temperature, smoke, and humidity) and can provide real-time data to building management systems. For instance, smoke detectors can send alerts to emergency services or building managers when a fire is detected.

- **Automated Systems:** Smart building systems can automate fire alarms, sprinkler systems, and emergency lighting. These systems can be programmed to respond to specific triggers, enhancing response times.
- **Data Analytics:** By analysing data collected from various sensors, building managers can identify patterns and predict potential fire hazards. This proactive approach allows for timely interventions before incidents escalate.
- **Remote Monitoring:** Facilities can be monitored remotely, allowing for immediate response to fire alerts. This is especially beneficial for large buildings or complexes where quick communication is crucial.
- **Integration with Other Systems:** Smart technologies can be integrated with HVAC and security systems to optimize responses during emergencies, such as adjusting airflow to contain smoke or providing access control during evacuation.

Fire Modelling Software and Simulations

Fire modelling software and simulations enable engineers and safety professionals to analyse and predict fire behaviour within buildings and other structures. These tools assist in designing fire safety measures and improving overall safety planning.

Key Components:

Computational Fluid Dynamics (CFD)

CFD modelling allows for the simulation of fire dynamics, including smoke movement, heat transfer, and combustion processes. This helps engineers understand how fires will behave under various conditions and informs design decisions.

Evacuation Modelling

Software such as Pathfinder and Mass Motion simulates occupant movement during a fire evacuation. These tools help identify potential bottlenecks and optimize building layouts for efficient evacuation routes.

Risk Assessment Tools

Fire modelling software can assess the risks associated with specific fire scenarios, considering factors like material properties and building configurations. This data is essential for regulatory compliance and insurance assessments.

Design Optimization

By using simulations, architects and engineers can refine building designs to incorporate effective fire safety features. This includes optimizing exit routes, selecting appropriate materials, and integrating fire suppression systems.

Future Trends in Fire Safety Solutions

The future of fire safety solutions is shaped by advancements in technology, increasing awareness of fire risks, and regulatory changes. Here are some notable trends:

AI and Machine Learning

AI can analyse vast amounts of data to improve fire detection systems, predict fire incidents, and enhance decision-making during emergencies. Machine learning algorithms can learn from past incidents and improve predictive capabilities over time.

Drones and Robotics

Drones equipped with thermal imaging cameras can assess fire incidents from the air, providing critical information to first responders. Robotics may also assist in fire suppression, allowing for safer operations in hazardous environments.

Augmented Reality (AR) and Virtual Reality (VR)

AR and VR technologies can enhance training for firefighters and emergency responders, simulating real-life fire scenarios in a controlled environment. This immersive training can improve response preparedness.

Smart Materials

The development of smart materials that can react to heat or fire is on the rise. These materials can change properties in response to temperature changes, potentially providing additional passive fire protection.

Sustainable Fire Safety Solutions

With the growing focus on sustainability, there is a trend toward developing eco-friendly fire safety products, such as environmentally friendly fire retardants and recyclable fire protection systems.

Integration with Smart Cities

As urban environments evolve into smart cities, fire safety systems will increasingly integrate with broader city management systems, enhancing coordination during emergencies and improving overall urban resilience.

Emerging technologies in fire engineering are reshaping how we approach fire safety. From smart building technologies that enhance real-time monitoring and response capabilities to sophisticated fire modelling software that aids in design and planning, these innovations are vital for improving safety outcomes. Looking ahead, advancements in AI, robotics, and sustainable materials promise to further enhance fire safety solutions, making buildings safer and more resilient against fire hazards.

Chapter 16

Fire Strategies

Fire strategies play a critical role in the built environment, ensuring safety, legal compliance, and risk mitigation in case of fire incidents. A fire engineer's perspective incorporates scientific principles, engineering methodologies, and regulatory requirements to develop comprehensive fire strategies for buildings. This document explores the fundamental components of fire strategies, their implementation, and key considerations in performance-based fire engineering.

Objectives of A Fire Strategy

A fire strategy aims to
- Protect life and ensure safe evacuation
- Prevent fire spread and structural collapse
- Minimize property and economic losses
- Ensure compliance with building regulations and fire codes
- Facilitate emergency response and firefighting operations
- Consider sustainability and resilience in fire safety measures

Components of A Fire Strategy

A fire strategy consists of various interdependent elements designed to ensure holistic fire safety. These include:

Means of Escape

Ensuring safe evacuation routes is paramount in any fire strategy. Considerations include:

Adequate number and width of exits

Travel distances to escape routes

Protected corridors and stairwells

Emergency lighting and signage

Evacuation strategies (e.g., simultaneous, phased, or stay-put policies)

Fire Detection and Alarm Systems

Fire engineers design detection and alarm systems based on:

Building occupancy and fire risks

Smoke and heat detectors placement

Alarm audibility and accessibility

Integration with emergency response systems

Maintenance and testing protocols

Passive Fire Protection (PFP)

Passive fire protection measures prevent fire spread through building compartmentalization and material selection:

- Fire-resistant walls, floors, and doors
- Fire-stopping around penetrations and service ducts
- Fire-resistant glazing and barriers
- Structural fire protection to maintain integrity

Active Fire Protection (AFP)

Active systems control or extinguish fires and include:

- Sprinkler systems (wet, dry, pre-action, or deluge)
- Gas suppression systems (e.g., CO_2, FM-200, inert gases)
- Fire hydrants and hose reels
- Automatic smoke ventilation and extraction systems

Firefighting Access and Facilities

Fire engineers ensure compliance with firefighter access requirements:

- Firefighting shafts and stairwells
- Dry and wet riser installations
- Firefighting lifts
- Access roads and hardstanding for fire appliances

Smoke Control and Ventilation

Smoke management strategies are critical in reducing casualties and property damage:

- Natural and mechanical smoke ventilation
- Pressurization systems for escape routes
- Smoke compartmentation and zoning

Human Behaviour and Evacuation Planning

Understanding occupant behaviour during fire incidents is vital for designing effective evacuation strategies:

- Occupancy load and vulnerability assessments
- Fire drills and staff training
- Communication systems for emergency management

Fire Strategy Development and Implementation

Regulatory Framework

Fire strategies must comply with national and international fire safety regulations and standards, such as:

- UK: Approved Document B (ADB), BS 9999, BS 7974
- USA: NFPA Codes and Standards, IBC, IFC
- EU: Eurocodes, EN 1991-1-2

Fire Risk Assessment

A fire risk assessment (FRA) is conducted to:

- Identify potential fire hazards
- Evaluate fire risks to occupants and property
- Recommend mitigation measures

Performance-Based Fire Engineering

Performance-based design (PBD) allows greater flexibility by applying fire engineering principles to:

- Simulate fire growth and spread
- Assess structural fire resistance using computational models
- Evaluate smoke movement and occupant evacuation using CFD simulations

Integration with Other Building Services

Fire strategies must align with:

Mechanical, electrical, and plumbing (MEP) systems

Energy efficiency and sustainability objectives

Smart building technologies (AI-driven fire detection and evacuation systems)

Challenges in Fire Strategy Implementation

- **Complex Building Designs**: High-rise buildings, mixed-use developments, and historical structures require tailored fire strategies.
- **Changing Regulations**: Compliance with evolving fire codes and best practices necessitates continuous adaptation.
- **Cost vs. Safety**: Balancing financial constraints with optimal fire safety measures is a common challenge.
- **Human Error and Maintenance**: Poor fire safety management, lack of training, and improper maintenance can compromise the effectiveness of fire strategies.

A well-developed fire strategy is essential for ensuring fire safety in the built environment. Fire engineers use a combination of prescriptive codes, performance-based design, and risk assessment methodologies to develop and implement effective fire strategies. Continuous review, testing, and adaptation of fire strategies in response to new technologies, regulatory updates, and lessons learned from fire incidents are crucial for enhancing safety and resilience.

The question is often asked "Should all fire strategies be developed by qualified fire engineers?" While this is not necessary in the most basic fire strategies a working knowledge of fire engineering is required for complicated buildings. It is important for one to recognise their limitations and not go beyond their competency level.

In the UK, Building Regulations, Regulation 38 requires the fire safety information of the building to be handed over to the owner on completion of the build.

Regulation 38 of The Building Regulations 2010

Regulation 38 of the Building Regulations 2010 (as amended) in England and Wales is a crucial requirement concerning fire safety information. It mandates that all relevant fire safety information be handed over to the responsible person (typically the building owner or manager) upon the completion of construction work on a building. This regulation ensures that the necessary information is available for the effective management of fire safety during the building's operational life.

Regulation 38 is particularly relevant for buildings where the Regulatory Reform (Fire Safety) Order 2005 (FSO) applies, meaning it impacts workplaces, public buildings, and multi-occupied residential premises.

Key Requirements of Regulation 38

Regulation 38 states:

"Where building work is carried out which affects fire safety, sufficient information must be provided to the responsible person to enable them to operate and maintain the building in compliance with fire safety law."

The regulation applies to:
- New buildings
- Material alterations to existing buildings*
- Changes of use that impact fire safety

It is the duty of the person carrying out the work (e.g., the developer, contractor, or building owner) to provide this fire safety information.

*Material Alterations to Existing Buildings

A material alteration of a building, as defined under the Building Regulations 2010 (England and Wales), refers to any work that significantly affects a building's structure, fire safety, or accessibility. Specifically, it applies to alterations that impact compliance with Regulation 3(2) of the Building Regulations.

Definition under Building Regulations

According to Regulation 3(2), an alteration is "material" if it affects compliance with the following key functional requirements:

Structural Safety (Part A)
- Any change that compromises the load-bearing capacity or stability of the building.
- Examples: Removing or altering load-bearing walls, modifying foundations, or adding new floors.

Fire Safety (Part B)
- Changes that worsen fire resistance, means of escape, fire detection, or containment of fire.
- Examples: Removing fire doors, altering fire compartmentation, modifying stairways, or blocking fire escape routes.

Accessibility (Part M)
- Alterations that reduce accessibility for disabled persons or do not comply with Part M.
  - Examples: Removing ramps, changing lift access, or modifying entrances without considering accessibility needs.

Other Key Safety-Related Requirements:

Combustion Appliances (Part J): Changes affecting ventilation or flue systems.

Protection from Falling (Part K): Removal or modification of balustrades, stairs, or guarding.

Electrical Safety (Part P): Changes affecting electrical circuits, particularly in dwellings.

Examples of Material Alterations

Some common alterations that would be classified as material include:

Fire Safety Changes:
- Replacing non-fire-rated doors with standard doors.
- Removing or modifying fire separation walls.
- Blocking or altering designated fire escape routes.
- Installing or removing a sprinkler system.

Structural Changes:
- Removing a chimney breast or load-bearing wall.
- Enlarging openings in structural walls.
- Converting a loft, basement, or garage.

Access and Use Modifications:
- Removing an accessible toilet or lift.
- Narrowing doorways or corridors in public buildings.
- Installing steps where a ramp was previously required.

Facade and External Alterations:
- Adding cladding or insulation that could affect fire safety.
- Changing external doors and windows that impact emergency escape routes.

What Information Must Be Provided?

The fire safety information provided under The Building Regulations, Regulation 38 should include, but is not limited to:

Fire Strategy Documents
- The overall fire strategy for the building, including fire prevention, detection, containment, and evacuation.
- Design principles that have been used to achieve fire safety compliance.

Plans and Layouts
- Fire compartmentation details (fire-resisting walls, doors, floors, and ceilings).

- Fire escape routes, exits, and access for emergency services.
- Location of fire protection systems (sprinklers, alarms, emergency lighting, etc.).
- Location of fire-fighting equipment and dry or wet risers.

Fire Safety Equipment and Systems

- Details of fire detection and alarm systems, including testing and maintenance schedules.
- Information on smoke control systems (natural and mechanical smoke ventilation).
- Fire suppression systems (e.g., sprinklers, gaseous suppression).
- Emergency lighting specifications.

Structural Fire Protection

Fire resistance ratings of materials and elements used in construction.

Fire stopping and penetration seals information.

Fire door specifications and maintenance instructions.

Management and Maintenance Information

- Responsibilities of the responsible person under the Regulatory Reform (Fire Safety) Order 2005.
- Maintenance schedules and testing procedures for fire safety systems.
- Fire safety risk assessment guidance.

Fire Service Access and Facilities

- Firefighter access routes and operational procedures.
- Water supply information for firefighting.
- Any specific fire service requirements included in the design.

Why is Regulation 38 Important?

- **Ensures Compliance with Fire Safety Legislation:** Helps responsible persons comply with the Regulatory Reform (Fire Safety) Order 2005.
- **Supports Fire Risk Assessments:** The information provided helps in conducting and maintaining an up-to-date fire risk assessment.
- **Improves Safety for Occupants:** Ensures that fire safety measures are understood and properly managed.
- **Facilitates Effective Emergency Response:** Helps firefighters and emergency services navigate and respond efficiently to fires.

Regulation 38 and Fire Safety in High-Risk Buildings

Following the Grenfell Tower fire in 2017, fire safety regulations have been under significant scrutiny. Regulation 38 has gained increased importance, particularly for high-rise and high-risk buildings. The Building Safety Act 2022 and Fire Safety Act 2021 reinforce the need for clear fire safety documentation, emphasizing the "golden thread" of information—ensuring fire safety details are accurately recorded and passed on throughout a building's lifecycle.

Regulation 38 of the Building Regulations 2010 is essential for ensuring that fire safety information is effectively transferred from those constructing buildings to those responsible for their management and maintenance. Compliance with this regulation helps ensure buildings are safe for occupants, aids in fire risk assessment, and supports emergency responders. Given the evolving landscape of fire safety regulations, it is crucial for developers, designers, and responsible persons to be fully aware of their obligations under Regulation 38.

A fire strategy will provide most, if not all of the documentation required for Regulation 38. The type of fire strategy is largely dependent upon the type and use of the building. For smaller low risk buildings, The Building Regulations Approved Document will suffice,

Approved Document B of The UK Building Regulations

Approved Document B (ADB) is a key part of the UK Building Regulations that provides guidance on fire safety in buildings. It is divided into two volumes:

Volume 1 – Dwellings (houses, flats, and other residential buildings)

Volume 2 – Buildings other than dwellings (commercial, public, and industrial buildings)

Purpose of Approved Document B

Approved Document B helps ensure that buildings are designed and constructed to minimize the risk of fire and protect occupants. It provides statutory guidance on meeting Part B (Fire Safety) of the Building Regulations in England.

Key Areas Covered

ADB provides guidance on:

Means of escape – Ensuring safe evacuation routes for occupants.

Fire spread – Controlling internal and external fire spread to protect buildings and neighbouring structures.

Structural fire resistance – Ensuring buildings can withstand fire for a set period.

Fire detection and warning systems – Installation of smoke alarms, fire alarms, and other warning systems.

Firefighting access and facilities – Providing access for fire services, including firefighting shafts and hydrants.

Changes and Updates

Approved Document B is periodically updated to improve fire safety standards. Some recent key changes include:

- **Ban on combustible materials** in external walls of high-rise residential buildings (following the Grenfell Tower fire).
- **Sprinkler requirements extended** to buildings over **11m high** (previously 18m).
- **Way finding signage for firefighters** in new buildings over 11m.
- **Mandatory fire safety plans** for complex buildings.

Relationship to Fire Engineering

While ADB provides prescriptive guidance, it allows for fire engineering solutions where alternative compliance methods can be justified. This is particularly relevant for complex buildings where standard solutions may not be suitable. Alternatives to ADB are BS9999 and BS9991.

BS 9999 provides a fire safety code of practice for the design, management, and use of buildings, integrating a risk-based approach to fire safety tailored to different building types and occupancies. It enhances flexibility in fire protection strategies by considering factors such as occupant characteristics, fire growth rates, and means of escape, allowing for more efficient design solutions compared to prescriptive standards.

BS 9991 focuses specifically on fire safety in residential buildings, including flats, houses, and specialized housing. It offers guidance on means of escape, fire detection and alarm systems, and compartmentation, with an emphasis on modern construction methods and fire engineering principles to improve fire protection while maintaining design flexibility.

BS 9999: Fire Safety in The Design, Management, and Use of Buildings

BS 9999 is a British Standard that provides a fire safety code of practice for the design, management, and use of buildings. It adopts a risk-based approach, offering fire engineers greater flexibility than traditional prescriptive fire safety regulations such as Approved Document B. By integrating occupant characteristics, fire growth rates, and means of escape strategies, BS 9999 enables fire engineers to tailor fire strategies to specific building types and uses while maintaining safety compliance.

Key Principles of BS 9999

BS 9999 replaces traditional prescriptive guidance with a structured risk-based methodology. It categorises buildings based on their occupancy characteristics, fire growth rates, and other risk factors to determine appropriate fire protection measures. The standard introduces:

Risk Profiling – A matrix system that categorises buildings based on:
- Occupancy characteristics (A, B, C, or D, depending on familiarity with the building and ability to evacuate independently).
- Fire growth rates (1 to 4, ranging from slow to ultra-fast fire growth).
- Management levels (1 to 3, reflecting the degree of fire safety management in place).

Means of Escape – Escape route dimensions and travel distances are determined using the risk profile, allowing for a more adaptable approach than rigid distance limits.

Compartmentation & Fire Resistance – Fire separation is based on risk levels rather than prescriptive wall and floor ratings, allowing for alternative design solutions through engineered approaches.

Smoke Control and Ventilation – Guidance on designing smoke control systems to assist with tenability conditions in escape routes and firefighting operations.

Firefighting Facilities – The standard details the provision of firefighting shafts, rising mains, and water supplies based on building height and occupancy risk.

Active and Passive Fire Protection Measures – BS 9999 aligns fire protection measures with the overall fire strategy, including detection and alarm systems, suppression systems, and materials selection.

Application to Fire Strategy Development

When developing a fire strategy for a building, fire engineers use BS 9999 to:

- **Assess Risk Profiles:**

Identifying the appropriate risk category for the building to determine suitable fire precautions.

- **Design Means of Escape:**

Using the standard's guidance on travel distances, stair widths, and evacuation strategies.

- **Determine Fire Resistance Requirements**:

Establishing compartmentation and structural fire protection levels based on risk assessment rather than arbitrary time ratings.

- **Specify Fire Detection & Alarm Systems**:

Recommending fire detection solutions suited to the risk profile and building management strategy.

- **Incorporate Fire Safety Management Principles**:

Ensuring that active and passive fire protection measures align with the building's intended use and occupant behaviour.

Advantages of Using BS 9999

- **Design Flexibility**: Enables a more tailored fire safety approach than rigid prescriptive standards.
- **Risk-Based Approach**: Allows for more proportional fire safety measures based on the actual risk rather than blanket regulations.
- **Alignment with Modern Construction Methods**: Supports innovative design solutions, including open-plan layouts, by providing engineered solutions.
- **Improved Evacuation Modelling**: Encourages the use of performance-based fire engineering to refine escape strategies.

For fire engineers, BS 9999 offers a comprehensive and adaptable framework for developing fire strategies that balance safety, cost-effectiveness, and architectural ambition. By leveraging its risk-based approach, fire engineers can design buildings that meet safety objectives while accommodating modern construction techniques and usage demands. The standard's flexibility, however, requires a thorough understanding of fire dynamics and building management to ensure that fire safety provisions remain robust and effective throughout a building's lifecycle.

BS 9991: Fire Safety in The Design, Management, And Use of Residential Buildings

BS 9991 is the British Standard that provides a fire safety code of practice for the design, management, and use of residential buildings, including flats, apartments, and specialised housing. It is a companion document to BS 9999 but focuses on residential occupancies, allowing for a more flexible, risk-based approach than the prescriptive guidance of Approved Document B (ADB). For fire engineers developing a fire strategy, BS 9991 provides key principles for ensuring compliance while allowing for innovative and performance-based fire safety solutions.

Key Principles of BS 9991

BS 9991 outlines fire safety strategies tailored to residential settings, with emphasis on occupant evacuation, fire protection measures, and management strategies. The core principles include:

Risk-Based Approach – Similar to BS 9999, BS 9991 incorporates a structured approach based on risk assessment rather than fixed prescriptive measures. It considers factors such as occupancy type, fire growth rates, and levels of fire safety management.

Means of Escape – The standard offers guidance on travel distances, escape route design, and stair capacity for various types of residential buildings, including single dwellings, flats, and maisonettes. It introduces:

- Use of extended travel distances with appropriate mitigation (e.g., automatic fire detection, suppression systems).
- Acceptance of single staircases in taller residential buildings, provided they are adequately protected with smoke ventilation systems.
- Alternative escape provisions for sheltered housing and specialist residential accommodation.

Fire Suppression Systems – BS 9991 strongly advocates the use of sprinkler systems in residential buildings, particularly in high-rise developments. This aligns with modern fire safety trends that prioritise active suppression to reduce fire spread and enhance occupant safety.

Smoke Control and Ventilation – The standard provides detailed recommendations on smoke ventilation systems, including:

- Natural and mechanical smoke ventilation for common corridors and stairwells.
- Considerations for firefighting shaft requirements in high-rise buildings.

Compartmentation and Fire Resistance – It specifies requirements for fire-resistant walls and floors, ensuring adequate separation between individual dwellings and common areas. The approach allows for reduced fire resistance in some scenarios where suppression systems and smoke ventilation are implemented.

Fire Detection and Alarm Systems – BS 9991 prescribes minimum standards for fire detection and alarm systems in different residential settings:

Grade D1/D2 smoke alarms for single dwellings.

Interlinked smoke and heat detection in flats and maisonettes.

Additional fire alarm measures for assisted living and care homes.

Firefighter Access and Facilities – The standard sets out requirements for firefighter stairwells, dry and wet rising mains, and fire-fighting shafts in taller buildings. It also provides recommendations on vehicle access and hydrant locations to support fire and rescue operations.

Application to Fire Strategy Development

When developing a fire strategy for a residential building, fire engineers use BS 9991 to:

- **Determine Appropriate Means of Escape**

Ensuring adequate provision of escape routes, protected staircases, and travel distances based on occupancy type and fire risk.

- **Specify Fire Suppression Systems**

Recommending sprinklers or misting systems as risk mitigation measures, especially in high-rise and vulnerable occupancies.

- **Assess Smoke Control Requirements**

Designing effective smoke ventilation systems to maintain escape route tenability.

- **Ensure Adequate Compartmentation**

Providing fire-resistant construction to prevent fire spread between dwellings and common areas.

- **Address Fire Detection and Alarm Systems**

Ensuring compliance with required levels of detection and interconnectivity.

- **Consider Firefighter Access**

Designing firefighting shafts, rising mains, and access routes in line with BS 9991 guidance.

Advantages of Using BS 9991

- **Increased Flexibility**: Allows performance-based design solutions that meet safety objectives while enabling architectural innovation.
- **Improved Fire Safety for Residential Buildings**: Encourages the use of suppression systems and modern smoke control measures to enhance occupant protection.
- **Alignment with Industry Trends**: Supports contemporary design approaches, such as open-plan layouts and single-staircase solutions, within an engineered fire strategy.
- **Risk-Based Fire Engineering**: Enables fire safety strategies to be proportionate to building use and occupancy needs.

BS 9991 provides fire engineers with a comprehensive framework for designing fire strategies in residential buildings. By applying its risk-based principles, engineers can ensure compliance while adopting modern construction techniques and improving overall fire safety. The standard's emphasis on suppression, smoke control, and tailored evacuation strategies makes it a vital tool for fire engineering in contemporary residential developments.

BS 7974: Application of Fire Safety Engineering Principles To The Design of Buildings

BS 7974 is the British Standard that provides a framework for the application of fire safety engineering (FSE) principles in the design of buildings. Unlike prescriptive guidance, BS 7974 enables fire engineers to develop performance-based fire strategies tailored to specific building designs and uses. This standard is particularly valuable for complex, innovative, or high-risk buildings where traditional design methods may be restrictive or impractical.

Key Principles of BS 7974

BS 7974 outlines a structured approach to fire engineering, allowing for greater flexibility in design while maintaining safety objectives. It consists of a core document supported by a series of Published Documents (PDs) that provide detailed guidance on specific aspects of fire engineering.

Fire Safety Engineering Framework

- BS 7974 defines a systematic approach to fire safety design, including hazard identification, fire scenario development, and assessment of fire safety measures.
- It emphasises an iterative design process, ensuring that fire safety objectives are met efficiently.

Fire Safety Design Process

- The fire engineering approach includes:
- Defining fire safety objectives.
- Identifying credible fire scenarios.
- Determining occupant behaviour and evacuation strategies.
- Assessing fire growth and smoke movement.
- Evaluating structural performance in fire conditions.

- Engineers use computational models and quantitative analysis to validate design decisions.

Published Documents Supporting BS 7974

- **PD 7974-1: Initiation and Development of Fire in Enclosures** – Covers fire dynamics and fire growth modelling.
- **PD 7974-2: Spread of Smoke and Toxic Gases** – Provides methodologies for assessing smoke movement and tenability conditions.
- **PD 7974-3: Structural Response and Fire Spread** – Examines fire resistance of structures and compartmentation effectiveness.
- **PD 7974-4: Detection, Activation, and Suppression** – Covers fire detection, alarm systems, and suppression technologies.
- **PD 7974-5: Human Factors and Evacuation** – Focuses on occupant behaviour and evacuation modelling.
- **PD 7974-6: Fire Services Intervention** – Provides guidance on designing buildings to support firefighting operations.
- **PD 7974-7: Probabilistic Risk Assessment** – Introduces risk-based methodologies for fire safety assessment.

Application to Fire Strategy Development

BS 7974 allows fire engineers to develop bespoke fire strategies based on a combination of scientific principles, risk assessment, and engineering analysis. Key applications include:

Performance-Based Design
- Utilising computational fluid dynamics (CFD) models to simulate fire and smoke spread.
- Designing alternative solutions where prescriptive guidance is restrictive (e.g., large open-plan spaces, complex buildings).

Evacuation Strategies
- Analysing occupant movement using evacuation modelling tools.
- Developing phased or progressive evacuation strategies in large or high-occupancy buildings.

Structural Fire Engineering
- Assessing fire resistance of structural elements under realistic fire conditions.
- Optimising material selection to balance cost, sustainability, and safety.

Smoke Control and Suppression Systems
- Designing mechanical or natural smoke ventilation systems.
- Assessing the effectiveness of sprinklers, water mist systems, and other suppression technologies.

Firefighting Provisions
- Ensuring adequate firefighter access and facilities.
- Designing buildings to facilitate fire and rescue service intervention.

Advantages of Using BS 7974

- **Flexibility and Innovation**

Encourages tailored fire strategies rather than rigid compliance with prescriptive codes.

- **Scientific Approach**

Uses fire dynamics, modelling, and risk assessment to create robust safety solutions.

- **Optimisation of Design and Cost**

Enables efficient use of space and materials while maintaining fire safety.

- **Alignment with Modern Building Techniques**

Supports contemporary construction methods, including mass timber and modular buildings.

- **Regulatory Compliance**

Provides a recognised framework that can be used to justify alternative fire safety solutions to authorities.

BS 7974 is an essential tool for fire engineers, offering a performance-based methodology for designing safe, efficient, and innovative buildings. By applying its principles, engineers can develop bespoke fire strategies that address complex challenges while ensuring regulatory compliance and occupant safety.

The appendix of this book contains templates for the various types of fire strategy using ADB, BS9999, BS9991 and BS7974.

The templates provide information to be included in the fire strategies, but this is not exhaustive, if using the templates your judgement should be used to adjust the content as and when necessary.

Chapter 17

Conclusion

Fire engineering is a discipline that bridges science, engineering, and human safety, with the goal of mitigating fire hazards and ensuring the protection of lives, property, and the environment. Throughout this book, we have explored the evolution of fire safety, the scientific principles governing fire behaviour, and the engineering strategies used to minimize risks. As buildings become taller, materials evolve, and technological advancements reshape the built environment, the importance of fire engineering has never been more crucial.

The study of fire dynamics has demonstrated the complexity of fire growth, heat transfer, smoke movement, and suppression mechanisms. Understanding how fire spreads, interacts with various materials, and responds to different environmental conditions is fundamental to designing effective fire protection systems. This book has detailed the role of both passive and active fire protection measures, emphasizing that fire safety is not a single solution but a comprehensive strategy that integrates multiple elements, from fire-resistant materials and compartmentation to advanced suppression and detection systems.

Regulatory frameworks and building codes have evolved in response to major fire disasters, and their implementation remains a cornerstone of fire safety. We have seen how historical incidents, such as the Great Fire of London, the Triangle Shirtwaist Factory Fire, and the Grenfell Tower Fire, have shaped today's fire safety policies. While fire codes and regulations vary across jurisdictions, the underlying principles of risk assessment, safety planning, and compliance with standards such as BS 7974, NFPA, and PAS 79 remain universally relevant.

This book emphasizes that fire safety is not just about engineering structures, but also about understanding human behaviour in emergencies. The psychology of evacuation, decision-making under pressure, and effective emergency communication all play critical roles in fire risk management. Fire drills, training programs, and public awareness campaigns are just as vital as advanced engineering solutions.

The Role of Technology and Future Challenges

As we move into the future, the role of emerging technologies in fire engineering is expanding rapidly. Computational fire modelling, artificial intelligence (AI), and the Internet of Things (IoT) are transforming how we predict, detect, and respond to fires. Smart buildings equipped with automated fire detection and suppression systems can provide real-time data, allowing fire safety professionals to make more informed decisions. Machine learning algorithms are improving fire risk assessment, while computational fluid dynamics (CFD) modelling enhances our ability to simulate fire behaviour and smoke movement.

However, with these advancements come new challenges. The rise of new materials, such as high-performance plastics, composite materials, and lithium-ion batteries, has introduced new fire risks that require innovative mitigation strategies. Fires involving renewable energy sources, such as solar panels and battery storage systems, pose unique hazards, particularly in terms of toxic smoke production, explosion risks, and suppression challenges. Fire engineers must continuously adapt to these changes, ensuring that safety measures evolve alongside technological progress.

The increasing development and density of cities also present significant challenges for fire safety. High-rise buildings, underground transport systems, and interconnected smart infrastructure require holistic fire engineering approaches that consider evacuation strategies, smoke control mechanisms, and fire service accessibility. Performance-based fire engineering design is becoming more prevalent, offering flexible solutions that go beyond prescriptive fire codes and regulations. However, these approaches demand a thorough understanding of fire science, risk assessment, and interdisciplinary collaboration.

The Importance of a Multi-Disciplinary Approach

Fire safety is not the responsibility of a single profession but rather a collaborative effort between fire engineers, architects, building surveyors, policymakers, and emergency services. The best fire protection strategies are those that integrate fire safety considerations from the earliest stages of design through to construction, operation, and ongoing maintenance. By fostering strong communication and shared expertise across disciplines, we can create safer built environments that are not only compliant with regulations but also proactively resilient to fire hazards.

Education and research in fire engineering continue to play a crucial role in advancing fire safety. Institutions, professional organizations, and fire safety associations provide the foundation for continuous learning, ensuring that fire engineers remain at the forefront of new developments in the field. Fire safety training and knowledge dissemination must extend beyond professionals—building occupants, business owners, and the general public all have a role to play in fire prevention and emergency preparedness.

Final Thoughts

This book has aimed to provide a comprehensive yet accessible guide to fire engineering, combining theoretical knowledge with practical applications. Whether you are a fire safety professional, engineer, building surveyor, regulatory authority, or student, the insights shared within these pages are intended to deepen your understanding of fire safety principles and empower you to implement best practices in the field.

While we have made significant progress in fire safety over the years, fire remains an unpredictable and evolving threat. It is our collective responsibility to continue innovating, learning, and collaborating to ensure that fire safety measures remain effective in a rapidly changing world. By applying the principles of fire science, safety engineering, and regulatory compliance, we can contribute to creating a future where fire-related tragedies are minimized, lives are protected, and communities are more resilient.

The journey of fire safety does not end here—it is an ongoing commitment to excellence, adaptability, and continuous improvement. Let this book serve as a resource, a guide, and an inspiration for those dedicated to advancing fire engineering and ensuring a safer world for all.

If you liked reading this book, look out for my next book,'*Performance Based Building Design*'.

Duncan Winsbury

Appendices

	Pg. No.
- Glossary of terms	229
- Resources for further reading	234
- Professional Organisations	235
- Fire Engineering Symbols and Formulas	239
- Example Reports	241
- PAS 79 Fire Risk Assessment	242
- Qualitative Fire Risk Assessment	246
- Semi Quantitative Fire Risk Assessment	250
- Quantitative Fire Risk Assessment	255
- Fire Strategy Approved Document B	259
- Fire Strategy BS9999	263
- Fire Zone Modelling	267
- Evacuation Modelling	270
- CFD Modelling	273
- Fire Investigation	276

Appendix 1

Glossary of Fire Engineering Terms

This glossary covers fundamental terms in fire engineering, providing a foundational understanding of the concepts and language used in the field.

A

- **Active Fire Protection:** Systems designed to extinguish or control a fire, such as sprinklers, fire alarms, and fire extinguishers.
- **AFFF (Aqueous Film-Forming Foam):** A type of firefighting foam that forms a film on the surface of flammable liquids to suppress fire.
- **Assembly Occupancy:** A type of occupancy where people gather for social, recreational, or entertainment purposes, such as theatres and stadiums.

B

- **Backdraft:** A rapid and explosive combustion that occurs when oxygen is suddenly introduced to a smouldering fire.
- **Breach:** A failure or compromise in the fire protection system that may allow fire or smoke to spread.
- **Burning Rate:** The speed at which a fuel burns, which affects the intensity and duration of a fire.

C

- **Compartmentation:** The practice of dividing a building into sections using fire-rated barriers to contain smoke and fire.
- **Combustible:** A material that can ignite and burn, as opposed to non-combustible materials that do not support combustion.
- **Conflagration:** A large, destructive fire that often involves multiple structures and poses a significant threat to life and property.

D

- **Detection System:** Devices and technologies used to detect the presence of fire or smoke, such as smoke detectors and heat detectors.

- **Dimensional Analysis:** A method of analysing fire behaviour using geometric and physical parameters.
- **Dry Chemical Extinguishers:** Fire extinguishers filled with a dry chemical agent that interrupts the chemical reaction of a fire.

E

- **Emergency Evacuation Plan:** A strategy outlining the procedures for evacuating a building in the event of a fire or other emergencies.
- **Engulfment:** The process in which a fire spreads rapidly and envelops an area or object.
- **Exit Route:** A designated path that occupants can take to safely exit a building during an emergency.

F

- **Fire Behaviour:** The way fire reacts to its environment, including factors like heat release, spread, and smoke production.
- **Fire Code:** Regulations established by local governments to ensure building safety and fire prevention.
- **Fire Dynamics:** The study of the physical and chemical processes involved in fire development, growth, and extinguishment.
- **Fire Resistance Rating:** The measure of how long a building material or assembly can withstand fire exposure without failing.

G

- **Ground Fire:** A fire that burns in or on the ground surface, including peat and other combustible materials.

H

- **Heat Release Rate (HRR):** The rate at which heat energy is released during combustion, a critical factor in fire dynamics.
- **Horizontal Exit:** An exit that leads occupants to a safe area on the same level, often used in larger buildings.

I

- **Ignition Source:** Any factor that can ignite combustible materials, including open flames, sparks, and hot surfaces.
- **Incident Command System (ICS):** A standardized approach to incident management that enables coordinated response to emergencies.

J

- **Jet Engine Fire:** A fire involving the fuel and combustion system of a jet engine, which can produce intense heat and flames.
- **Junction Box:** An enclosure for electrical connections, where wires from different circuits meet, which can be a fire hazard if not properly installed.

K

- **Knockdown:** The rapid suppression of a fire, typically referring to the initial successful efforts to control or extinguish a fire.
- **K Class Fire:** A classification of fires involving cooking oils and fats, often requiring specialized extinguishing agents like wet chemical extinguishers.

L

- **Life Safety Systems:** Systems designed to protect the life and safety of occupants during a fire, including alarms, exit signs, and smoke control systems.
- **Low-Visibility Conditions:** Situations in which visibility is significantly reduced due to smoke or other factors during a fire.

M

- **Manual Fire Alarm System:** A fire alarm system that requires human intervention to activate, such as pulling a lever or pushing a button.
- **Material Safety Data Sheet (MSDS):** A document that provides information on the properties, hazards, and handling of specific substances, including fire risks.

N

- **NFPA (National Fire Protection Association):** An organization that develops and publishes fire safety codes and standards in the United States to promote fire prevention and protection.
- **Nozzle:** A device attached to a fire hose that directs the flow of water or firefighting foam and controls the shape and pattern of the spray.
- **Non-combustible Material:** A material that does not ignite or burn in a fire, often used in construction to enhance fire resistance.
- **Natural Ventilation:** The process of allowing outside air to enter a building to control temperature and smoke movement during a fire without mechanical systems.

O

- **Occupant Load:** The total number of people expected to occupy a building or space, which influences fire safety measures and evacuation planning.

- **Overhaul:** The process of checking for hidden fires after the main fire has been extinguished to ensure it is completely out.

P

- **Passive Fire Protection:** Measures designed to contain fires or slow their spread without requiring mechanical or electrical systems, such as fire-rated walls and doors.
- **Pyrolysis:** The thermal decomposition of materials at elevated temperatures in the absence of oxygen, which can lead to fire.
- **Pre-Incident Planning:** The process of developing strategies and procedures for responding to potential fire incidents in specific buildings or areas.

Q

- **Quenching:** The rapid cooling of a material, often after it has been heated to high temperatures, which can influence fire dynamics and behaviour.
- **Quality Assurance:** Procedures and practices used to ensure that fire protection systems and materials meet specified standards and regulations.

R

- **Rapid Fire Spread:** A situation in which a fire quickly expands, often influenced by available fuels and ventilation conditions.
- **Retardant:** A substance used to slow down or inhibit the spread of fire, commonly applied in firefighting efforts.

S

- **Smoke Control System:** Systems designed to manage and mitigate smoke movement during a fire, improving visibility and reducing toxic exposure.
- **Sprinkler System:** A fire protection system that uses a network of pipes and sprinkler heads to distribute water in the event of a fire.

T

- **Thermal Imaging Camera:** A device that detects heat and visualizes it, often used by firefighters to locate hotspots and victims in smoke-filled environments.
- **Total Flooding System:** A fire suppression system that fills an enclosed space with extinguishing agent, displacing oxygen to suppress fire.

U

- Unobstructed Exit: An exit that is clear of any obstacles, allowing for safe egress during an emergency.

V

- **Ventilation:** The process of controlling air flow in a building to manage smoke and heat during a fire.
- **Vapor Pressure:** The pressure exerted by a vapor in equilibrium with its liquid or solid form, influencing flammability and fire hazards.

W

- **Water Mist System:** A fire suppression system that uses fine water droplets to cool the fire and displace oxygen.
- **Wildfire:** An uncontrolled fire that burns in natural areas such as forests, grasslands, or prairies, often influenced by environmental conditions.

X

- **X-ray Inspection:** A technique used to examine the integrity of fire barriers, ducts

Y

- **Yield Point:** The stress level at which a material begins to deform permanently; important in evaluating the structural integrity of materials in fire conditions.
- **Yoke:** A structural component in some fire protection systems, particularly in suspended ceilings, that helps support and stabilize fire suppression equipment.

Z

- **Zone:** A designated area within a building or site, often used to describe specific sections for fire safety systems, such as alarm zones or smoke control zones.
- **Zoning:** The practice of dividing a building into various areas or zones for fire safety management, allowing for tailored fire protection measures and effective evacuation plans.

Appendix 2
Resources for Further Reading in Fire Engineering Textbooks

	Author / Publisher	Title	Remarks
1	James L. Pharr	Fundamentals of Fire Protection Engineering	A comprehensive introduction to fire protection engineering principles and practices.
2	Gregory E. Gorbett and James L. Pharr	Fire Dynamics	This book provides an in-depth exploration of the physics of fire behaviour and the factors affecting it.
3	National Fire Protection Association (NFPA)	Fire Protection Handbook	A comprehensive reference that covers a wide range of fire safety topics, including design, engineering, and fire safety systems.
4	Joseph M. T. A. D. Sorensen	Principles of Fire Protection	A textbook covering essential fire safety principles, engineering practices, and regulatory standards.
5	Robert W. Fitzgerald and William A. P. M. Schaefer	Introduction to Fire Protection	A foundational textbook on the principles and practices of fire protection engineering.

Appendix 3
Professional Organizations
UK Professional Organizations for Fire Engineering

	Organization	Website	Overview
1	Institution of Fire Engineers (IFE)	www.ife.org.uk	A leading professional body for fire engineers and fire safety professionals, offering membership, training, and certification. The IFE promotes fire engineering as a profession and supports the continuous professional development of its members.
2	Fire Protection Association (FPA)	www.thefpa.co.uk	The UK's national fire safety organization, providing guidance, training, and resources to help organizations and individuals improve fire safety practices and comply with regulations.
3	National Fire Chiefs Council (NFCC)	www.nationalfirechiefs.org.uk	An organization representing fire and rescue services in the UK, focusing on improving fire safety and responding to fires and emergencies through collaboration and knowledge sharing.
4	Building Research Establishment (BRE)	www.bregroup.com	A leading research organization that provides fire safety engineering services, certifications, and guidance on building and fire safety standards.
5	UK Fire Safety Engineering Group (UKFSEG)	www.ukfseg.org.uk	A group of professionals dedicated to advancing the science and practice of fire safety engineering through knowledge sharing and collaboration.

	Organization	Website	Overview
6	Royal Institute of British Architects (RIBA)	www.architecture.com	While primarily focused on architecture, RIBA provides resources and guidance on fire safety and regulations relevant to the built environment.
7	Chartered Institute of Building (CIOB)	www.ciob.org	An organization representing construction professionals that also emphasizes the importance of fire safety in construction projects.
8	Institution of Mechanical Engineers (IMechE) - Fire Engineering Group	www.imeche.org	The IMechE includes a fire engineering group that focuses on the mechanical aspects of fire safety engineering and technology.
9	Society of Fire Protection Engineers (SFPE) - UK Chapter	www.sfpe.org	The UK chapter of the international SFPE, promoting fire protection engineering through education, research, and advocacy.
10	The Fire Safety Engineering Group (FSEG) at the University of Greenwich	www.gre.ac.uk	This academic group focuses on research, education, and consultancy in fire safety engineering, contributing to industry standards and best practices.
			These organizations play a significant role in promoting fire engineering and safety in the UK, offering resources, networking opportunities, and professional development for individuals in the field. Whether you are looking for training, certification, or collaboration, these organizations provide valuable support to fire engineering professionals.

	Organization	Website	Overview
11	Institute of Fire Safety Managers (IFSM)	www.ifsm.org.uk	The IFSM is a professional body for fire safety managers and professionals working in fire safety management. It was established to support the development and recognition of fire safety management as a vital discipline within the broader field of fire safety. **Key Features and Activities:** - **Membership:** The IFSM offers various membership levels, providing opportunities for professionals at different stages of their careers. Members benefit from networking, professional development, and access to resources. - **Professional Development:** The organization provides training, seminars, and workshops focused on current fire safety practices, legislation, and industry standards. - **Certification:** The IFSM offers a certification program for fire safety managers, recognizing individuals who demonstrate competence in fire safety management principles and practices.

Organization	Website	Overview
		• **Guidance and Resources:** The IFSM publishes best practice guides, research papers, and other resources to help fire safety managers effectively implement fire safety measures and comply with regulations. • **Advocacy:** The institute advocates for fire safety professionals and works to influence fire safety legislation and policy at national and local levels. • **Networking Opportunities:** The IFSM organizes events, conferences, and forums where members can connect, share knowledge, and discuss industry challenges and developments.

Apendix 4

Fire Engineering Symbols and Formulas

Symbol	Meaning
Q	Heat Release Rate (kW)
T	Temperature (°C or K)
ρ	Density (kg/m³)
h	Height (m)
A	Area (m²)
V	Volume (m³)
$\dot{m}$	Mass Flow Rate (kg/s)
U	Velocity (m/s)
c_p	Specific Heat Capacity (J/kg·K)
k	Thermal Conductivity (W/m·K)
g	Acceleration due to Gravity (m/s²)
t	Time (s)
P	Pressure (Pa)
I	Radiative Intensity (W/m²)
F	View Factor (-)
α	Absorptivity (-)
ε	Emissivity (-)

General Formulas
Greek Alphabet

Capital	Small	Name
A	α	Alpha
B	β	Beta
Γ	γ	Gamma

Δ	δ	Delta
Ε	ε	Epsilon
Ζ	ζ	Zeta
Η	η	Eta
Θ	θ	Theta
Ι	ι	Iota
Κ	κ	Kappa
Λ	λ	Lambda
Μ	μ	Mu
Ν	ν	Nu
Ξ	ξ	Xi
Ο	ο	Omicron
Π	π	Pi
Ρ	ρ	Rho
Σ	σ / ς	Sigma
Τ	τ	Tau
Υ	υ	Upsilon
Φ	φ	Phi
Χ	χ	Chi
Ψ	ψ	Psi
Ω	ω	Omega

Greek Letters in Fire Engineering Science

Symbol	Parameter	Unit
α	Thermal Diffusivity	m²/s
β	Coefficient of Thermal Expansion	1/K
γ	Surface Tension	N/m
δ	Boundary Layer Thickness	m
ε	Emissivity	-
ζ	Flame Spreading Factor	-

η	Efficiency	-
θ	Temperature Difference	K
ι	Radiative Transfer Coefficient	W/m^2K
κ	Thermal Conductivity	$W/m \cdot K$
λ	Wavelength	m
μ	Dynamic Viscosity	$Pa \cdot s$
ν	Kinematic Viscosity	m^2/s
ξ	Flame Tilt Angle	°
ο	Solid Angle	sr
π	Mathematical Constant (Pi)	-
ρ	Density	kg/m^3
σ	Stefan-Boltzmann Constant	W/m^2K^4
τ	Shear Stress	Pa
φ	Relative Humidity	%
χ	Mole Fraction	-
ψ	Humidity Ratio	kg/kg
ω	Angular Velocity	rad/s

Appendix 5
Example Reports

- PAS 79 Fire Risk Assessment
- Qualitative Fire Risk Assessment
- Semi Quantitative Fire Risk Assessment
- Quantitative Fire Risk Assessment
- Fire Strategy Approved Document B
- Fire Strategy BS9999
- Fire Zone Modelling
- Evacuation Modelling
- CFD Modelling
- Fire Investigation

While every care has been taken to ensure that the content of the example reports meet the relevant requirements, it is the responsibility of the author of the reports to ensure the content contains the required information for its intended use.

Pas 79 Fire Risk Assessment Template

Pas 79 Fire Risk Assessment

1. General Information

Item	Details
Premises Name	
Address	
Assessor's Name	
Date of Assessment	
Responsible Person	
Contact Information	
Type of Premises	
Occupancy Profile	
Number of Floors	
Approximate Floor Area	

2. Legislative Compliance

This fire risk assessment has been conducted in accordance with **PAS 79:2020** and the **Regulatory Reform (Fire Safety) Order 2005**. It follows the methodology recommended by the **British Standards Institution (BSI)** to ensure a structured and systematic approach to fire risk assessment.

3. Fire Hazards And Prevention Measures

Fire Hazard	Risk Level (Low/ Medium/High)	Existing Control Measures	Additional Actions Required
Sources of Ignition			

Fire Hazard	Risk Level (Low/Medium/High)	Existing Control Measures	Additional Actions Required
Sources of Fuel			
Oxygen Availability			
Housekeeping			
Waste Management			

4. People At Risk

Category	At Risk (Yes/No)	Notes
Employees		
Visitors		
Contractors		
Residents/Tenants		
Disabled Individuals		
Other Vulnerable Groups		

5. Fire Protection Measures

Fire Protection Feature	Available (Yes/No)	Condition	Notes
Fire Detection System			
Fire Alarm System			
Emergency Lighting			
Fire Extinguishers			
Sprinkler System			
Smoke Control Systems			

Fire Protection Feature	Available (Yes/No)	Condition	Notes
Fire Doors & Compartmentation			

6. Means of Escape and Evacuation Procedures

Escape Route Element	Condition (Good/Fair/Poor)	Notes
Escape Routes Clear & Unobstructed		
Signage & Wayfinding		
Emergency Exit Doors		
Assembly Points		
Evacuation Plans Available		
Fire Drills Conducted		

7. Emergency Planning and Training

Aspect	Yes/No	Notes
Fire Emergency Plan Exists		
Fire Drills Conducted		
Fire Warden Appointed		
Fire Safety Training Given		
First Aid Arrangements		

8. Significant Findings & Action Plan

Finding	Risk Level (L/M/H)	Recommended Action	Responsible Person	Target Completion Date

9. Assessor's Declaration

I confirm that this fire risk assessment has been conducted in accordance with PAS 79 methodology and the **Regulatory Reform (Fire Safety) Order 2005**. The findings are based on a site inspection and the information provided.

Assessor's Name	Signature	Date
Validator's Name	Signature	Date

This **Fire Risk Assessment** should be reviewed regularly and updated in response to changes in premises, processes, or occupancy.

Qualitative Fire Risk Assessment Template
Qualitative Fire Risk Assessment

1. General Information

Item	Details
Premises Name	
Address	
Assessor's Name	
Assessment Date	
Next Review Date	
Purpose of Assessment	

2. Description of Premises

Item	Details
Type of Premises	(e.g., office, retail store, warehouse, residential building)
Number of Floors	
Occupancy Details	(e.g., number of occupants, staff, visitors, vulnerable individuals)
Key Activities	

3. Fire Hazard Identification

Hazard Type	Description
Ignition Sources	(e.g., electrical equipment, cooking appliances, smoking materials, hot works)
Fuel Sources	(e.g., paper, wood, textiles, chemicals, furniture)
Oxygen Supply	(e.g., natural ventilation, air conditioning, open doors/windows)

4. People at Risk

Category	Description
Employees	(including lone workers)
Visitors and Customers	
Residents (if applicable)	
Vulnerable Individuals	(e.g., elderly, disabled persons, children)

5. Existing Fire Safety Measures

Safety Measure	Details
Fire Detection and Warning Systems	(e.g., smoke detectors, manual call points, alarms)
Firefighting Equipment	(e.g., fire extinguishers, fire blankets, sprinkler systems)
Emergency Exits and Escape Routes	(e.g., number, signage, accessibility, lighting)
Fire Safety Signage and Notices	
Fire Safety Training and Drills	
Housekeeping and Storage Practices	

6. Fire Risk Evaluation

Factor	Rating (Low / Medium / High)
Likelihood of Fire Occurring	
Potential Consequences	
Overall Fire Risk Rating	

7. Recommended Improvements

Improvement Area	Action Required	Responsible Person	Target Completion Date
Housekeeping and Waste Management			

Improvement Area	Action Required	Responsible Person	Target Completion Date
Fire Detection System Upgrade			
Staff Training and Fire Drills			
Signage and Escape Route Enhancement			
Electrical Equipment Maintenance			
Other (specify)			

8. Documentation and Review

Item	Details
Assessment Completed By	
Signature	
Date	
Next Review Date	

9. Regulatory Reform (Fire Safety) Order 2005

The **Regulatory Reform (Fire Safety) Order 2005** (RRO) is the primary legislation governing fire safety in non-domestic premises in England and Wales. It requires that:

- A **responsible person** (e.g., employer, owner, occupier) must ensure fire safety in the premises.
- A **fire risk assessment** is conducted and regularly reviewed.
- Adequate **fire safety measures** are implemented, including detection systems, emergency exits, and firefighting equipment.
- Fire safety training is provided to staff and occupants.
- Fire prevention measures are maintained to reduce risk.

The **responsible person** must take reasonable steps to reduce the likelihood of fire, ensure safe evacuation procedures, and comply with fire safety regulations. Non-compliance can lead to enforcement action, fines, or prosecution by the fire authorities.

For more details, refer to official guidance provided by the UK Government.

Notes:
- This qualitative fire risk assessment is based on subjective expert judgment and observations.
- Regular reviews and updates are essential to ensure ongoing fire safety compliance.

Semi-Quantitative Fire Risk Assessment (SQFRA) Template
Semi-Quantitative Fire Risk Assessment

1. General Information

1.1. Premises Details

- Building Name:
- Address:
- Type of Premises (e.g., commercial office, retail space):
- Occupancy Type:
- Number of Occupants:
- Assessment Date:
- Assessor Name:
- Assessor Qualification/Position:

1.2. Scope of the Assessment

- Purpose of the Assessment: (e.g., regulatory compliance, insurance requirements, safety improvement)
- Areas Assessed: (e.g., entire building, specific floors, high-risk zones)
- Assessment Methodology: (Numerical risk scoring system)

2. Fire Hazards Identification

2.1. Ignition Sources *(List and rate likelihood of each source contributing to a fire incident)*

Ignition Source	Present? (Yes/No)	Likelihood Rating (1-5)
Electrical faults (overloaded circuits, faulty wiring)		
Open flames (candles, stoves)		
Heating equipment (boilers, radiators)		
Smoking materials		
Hot work (welding, cutting, grinding)		
Others (specify)		

2.2. Fuel Sources *(Rate the combustibility of materials present in the building)*

Fuel Source	Present? (Yes/No)	Combustibility Rating (1-5)
Paper, cardboard, and office materials		
Flammable liquids (paints, solvents)		
Furniture and upholstery		
Plastics and synthetics		
Gas supplies		
Others (specify)		

2.3. Oxygen Supply (Ventilation & Airflow)

- Ventilation Type: (Natural / Mechanical)
- Does ventilation contribute to fire spread? (Yes/No)
- Air Conditioning/Ducting Present? (Yes/No)

3. Fire Risk Scoring System

3.1. Likelihood of Fire Occurrence

Rate the likelihood of fire occurring based on ignition sources, fuel load, and human factors.

Likelihood Level	Score	Description
Very Unlikely	1	Rare event, no known history
Unlikely	2	Possible but not expected
Possible	3	Could happen occasionally
Likely	4	Happens regularly
Very Likely	5	Frequent occurrence or near misses reported

3.2. Consequence Severity Rating

Rate the potential impact if a fire occurs.

Consequence Level	Score	Description
Insignificant	1	No injuries, minor damage
Minor	2	Small localized fire, no significant injuries
Moderate	3	Fire controlled, some damage, minor injuries
Major	4	Fire spreads, serious injuries, structural damage
Catastrophic	5	Large-scale fire, fatalities, building loss

3.3. Fire Risk Calculation

Use the formula:

Fire Risk Score = Likelihood Score × Consequence Score

Fire Risk Score	Risk Level	Action Required
1-4	Low	Acceptable risk, monitor
5-9	Medium	Implement risk reduction measures
10-16	High	Immediate action required
17-25	Critical	Urgent intervention needed

4. Fire Protection & Control Measures

4.1. Fire Detection & Alarm Systems

Feature	Present? (Yes/No)	Condition (Good/Poor)	Notes
Smoke detectors			
Heat detectors			
Automatic fire alarm system			
Manual call points			
Others			

4.2. Fire Suppression Systems

Suppression System	Present? (Yes/No)	Condition (Good/Poor)	Notes
Fire extinguishers			
Fire blankets			
Sprinkler systems			
Hose reels			
Others			

4.3. Means of Escape & Emergency Planning

- **Number of Exits Available:**
- **Exit Signage Condition:** (Good / Poor / Needs Improvement)
- **Emergency Lighting Condition:** (Good / Poor / Needs Improvement)
- **Fire Drills Conducted?** (Yes/No, frequency)
- **Evacuation Plan Available?** (Yes/No)
- **Fire Safety Training for Staff?** (Yes/No, last date)

5. Risk Mitigation & Action Plan

Based on the fire risk rating, outline corrective actions.

Identified Risk	Risk Score	Recommended Action	Priority (High/Medium/Low)	Responsible Person	Deadline
Example: Electrical faults	12 (High)	Regular PAT testing and circuit inspections	High	Facility Manager	1 month
Example: Blocked escape route	9 (Medium)	Remove obstructions and conduct checks	Medium	Safety Officer	2 weeks
Example: Fire extinguisher missing	6 (Medium)	Install new extinguisher	Medium	Maintenance	1 week

6. Summary & Conclusion

Overall Fire Risk Level: *(Low/Medium/High/Critical based on highest risk rating)*
Recommended Actions Summary: *(Brief description of key risks and mitigation steps)*

Final Comments: *(Any additional notes on fire safety improvements or concerns)*

Assessor Signature: _____

Date: _____

Quantitative Fire Risk Assessment (QFRA) Template

Quantitative Fire Risk Assessment

1. Introduction

1.1 Purpose of the Assessment

Provide an overview of the fire risk assessment objectives, including the scope, intended outcomes, and applicable regulations or standards.

Name of Author	Qualifications	Date
Name of Validator	**Qualifications**	**Date**

1.2 Scope

Define the boundaries of the assessment, including:
- Building type and function
- Fire hazards and risks considered
- Fire protection systems included
- Occupant characteristics and evacuation capabilities

2. Data Collection

2.1 Fire Incident Data

Source	Data Type	Relevance
Historical fire incident reports	Fire frequency, causes	Establish baseline probabilities
Industry statistics	Failure rates of fire protection systems	Reliability analysis
Experimental studies	Fire growth rates, toxicity	Model validation

2.2 System Performance Data

System	Performance Parameter	Data Source
Sprinkler system	Activation reliability	Manufacturer specifications, fire test data
Smoke detectors	False alarm rate, detection time	Fire test data, reliability databases
Passive fire protection	Fire resistance rating	Construction codes, material testing

3. Fire Scenario Modelling

3.1 Identification of Fire Scenarios

Develop representative fire scenarios based on:
- Common ignition sources
- Likely fuel loads
- Fire protection system response
- Occupant response time

Scenario ID	Ignition Source	Fuel Load	Fire Protection Response	Expected Outcomes
S1	Electrical fault	Office furniture	Sprinklers activate in 90s	Fire controlled, minor damage
S2	Cooking fire	Commercial kitchen	Suppression system fails	Rapid spread, full evacuation needed

3.2 Fire Growth Modelling

Use fire growth models such as:
- t-squared fire growth curves
- Computational Fluid Dynamics (CFD) simulations
- Zone models (e.g., B-Risk)

3.3 Evacuation Modelling

Utilize evacuation models such as:
- Pathfinder, STEPS, or EXODUS

- Time-based egress simulations
- Human behaviour analysis

4. Probabilistic Risk Analysis

4.1 Frequency Analysis

Apply statistical methods to determine fire occurrence probabilities:

Method	Application
Monte Carlo Simulations	Estimating fire spread outcomes
Fault Tree Analysis (FTA)	Identifying failure probabilities of fire protection systems
Event Tree Analysis (ETA)	Evaluating possible fire scenarios and outcomes

4.2 Consequence Analysis

Quantify fire impact in terms of:

- Property damage (£ value)
- Injury/fatality probabilities
- Business interruption costs

Scenario	Damage Estimate	Occupant Impact	Recovery Time
S1	£50,000	No casualties	1 week
S2	£1,000,000	Multiple injuries	6 months

5. Risk Evaluation and Mitigation

5.1 Risk Criteria

Establish risk acceptance criteria based on:

- ALARP (As Low As Reasonably Practicable) principle
- Regulatory standards
- Corporate risk thresholds

5.2 Risk Reduction Measures

Measure	Expected Risk Reduction	Cost Consideration
Enhanced sprinkler coverage	60% reduction in fire spread risk	Medium
Improved detection system	40% reduction in detection time	Low
Fire safety training	Faster occupant response, reduced injuries	Low

6. Validation and Sensitivity Analysis

6.1 Validation Against Real-World Data

Compare model results with:
- Historical fire incidents
- Full-scale fire test data
- Post-fire investigations

6.2 Sensitivity Analysis

Determine how uncertainties in inputs affect risk assessment outcomes.

Parameter	Sensitivity Impact
Detection time variation	High impact on evacuation success
Sprinkler activation failure	Moderate impact on fire spread

7. Conclusion and Recommendations

Summarize key findings, highlight high-risk areas, and recommend improvements. Provide a structured plan for risk mitigation and future reassessment.

8. References and Appendices

List all data sources, models, and assumptions used. Include supplementary materials such as detailed calculations, regulatory requirements, and technical specifications.

Fire Strategy Template Based on Approved Document B (ADB)

Fire Strategy Report

Prepared by: [Fire Engineer Name]
Company: [Fire Consultancy Name]
Date: [Insert Date]
Peer Reviewed by: [Fire Engineer Name]
Company: [Fire Consultancy Name]
Date: [Insert Date]
Version: [Insert Version Number]

1. Introduction

1.1 Purpose of the Fire Strategy

This fire strategy has been developed in accordance with **Approved Document B (ADB)** of the Building Regulations to ensure the safety of occupants and compliance with UK fire safety legislation. It outlines the fire safety provisions for [Building Name], ensuring appropriate means of escape, fire protection measures, and firefighting provisions.

1.2 Building Description

Building Name: [Insert Name]
- **Building Address**: [Insert Address]
- **Building Type**: [Residential / Commercial / Industrial / Mixed-Use]
- **Use Class**: [Insert Use Class]
- **Number of Storeys**: [Insert Number]
- **Total Floor Area**: [Insert Area in m²]
- **Occupant Load**: [Insert Number]
- **Construction Type**: [Insert Type, e.g., Steel Frame, Timber Frame, Concrete]

2. Means of Escape

2.1 Escape Routes and Travel Distances

- Escape routes designed in accordance with **ADB Volume 1 (Dwellings) or Volume 2 (Buildings other than dwellings)**.
- Number of Escape Stairs

- Maximum travel distances:
- **Single direction of travel**: [Insert Distance]m (ADB Limit: [Insert Limit])
- **Two directions of travel**: [Insert Distance]m (ADB Limit: [Insert Limit])
- Doors leading to escape routes to provide at least **FD30S/FD60S fire resistance** as required.

2.2 Final Exits and Assembly Points

- Final exits provided at ground level leading to a place of ultimate safety.
- Clear, unobstructed egress routes leading to designated assembly points **[Insert Location]**.

2.3 Refuge Areas for Disabled Occupants

- Refuge points located at **[Insert Locations]** with emergency communication points.
- Lifts not to be used for evacuation unless specifically designed as evacuation lifts.

3. Fire Compartmentation and Passive Fire Protection

3.1 Compartmentation Strategy

- Fire compartments designed to limit fire spread in accordance with **ADB Table B3**.
- **Walls and floors** to have a minimum fire resistance of **[Insert Duration]**.
- **Fire doors** to be installed in line with **ADB Table C1**.

3.2 Fire Stopping and Cavity Barriers

- Fire stopping to be provided at all **penetrations and junctions**.
- Cavity barriers installed in voids to prevent unseen fire spread.

4. Fire Detection and Alarm Systems

- Fire alarm system designed in accordance with **BS 5839-1 (Commercial) or BS 5839-6 (Residential)**.
- **Category and Grade**: [Insert System Type, e.g., L2, L3, Grade D2].
- Smoke and heat detection installed in all high-risk areas.

5. Fire Suppression Systems

- **Sprinkler System**: Required for residential buildings over 11m in height in accordance with **ADB Volume 1**.

System Compliance: Designed to **BS 9251 (Residential) / BS EN 12845 (Commercial)**.

Alternative suppression systems: [Insert Type, e.g., Water Mist, Gas Suppression].

6. Smoke Control and Ventilation

- **Corridor smoke ventilation**: [Natural / Mechanical] as per **ADB Section 3.6**.
- **Stairwell pressurisation system** installed where required.
- Smoke shafts sized to meet **ADB and BS 9999** requirements.

7. Firefighting Access and Facilities

7.1 Fire Appliance Access

- Firefighting access designed in accordance with **ADB Section 15**.
- Minimum **3.7m road width** provided for fire appliances.
- Fire service vehicle access provided to within **45m** of any dwelling entrance.

7.2 Fire Hydrants and Water Supply

- Hydrant locations in accordance with **ADB and BS 9990**.
- Flow rate of **[Insert L/min]** maintained.

7.3 Firefighting Shafts and Lifts

- Firefighting shafts provided in accordance with **ADB Table B5**.
- Firefighting lifts where required for buildings above **18m**.

8. Management and Maintenance

8.1 Fire Safety Management Plan

- Regular **fire drills and staff training** to be conducted.
- Fire doors, detection systems, and suppression equipment to be inspected in accordance with **BS 9999**.
- Fire risk assessments carried out under the **Regulatory Reform (Fire Safety) Order 2005**.

9. Conclusion

This fire strategy demonstrates compliance with **Approved Document B**, ensuring adequate fire protection, safe means of escape, and firefighting facilities. It provides a robust approach to fire safety tailored to the building's design and usage. Any deviations from prescriptive guidance are justified through fire engineering principles.

Fire Strategy Template Based on BS 9999
Fire Strategy Report

Prepared by: [Fire Engineer Name]

Company: [Fire Consultancy Name]

Date: [Insert Date]
Peer Reviewed by: [Fire Engineer Name]
Company: [Fire Consultancy Name]
Date: [Insert Date]

Version: [Insert Version Number]

1. Introduction

1.1 Purpose of the Fire Strategy

This fire strategy has been developed in accordance with **BS 9999: Code of Practice for Fire Safety in the Design, Management, and Use of Buildings**. It provides a risk-based approach to fire safety, considering occupancy, fire growth rates, and management levels to ensure a performance-based fire safety design for [Building Name].

1.2 Building Description

Building Name: [Insert Name]

Building Address: [Insert Address]

Building Type: [Commercial / Residential / Industrial / Mixed-Use]

Use Class: [Insert Use Class]

Number of Storeys: [Insert Number]

Number of Escape Stairs: [Insert Number]

Total Floor Area: [Insert Area in m²]

Occupant Load: [Insert Number]

Construction Type: [Insert Type, e.g., Steel Frame, Concrete, Timber Frame]

2. Risk Profile Assessment

BS 9999 applies a risk profile system to determine appropriate fire safety measures. This is based on:

- **Occupant Characteristics (A-D)**: [Insert Category]
- **Fire Growth Rate (1-4)**: [Insert Category]
- **Fire Safety Management Level (1-3)**: [Insert Category]

The combination of these factors defines the appropriate fire safety measures for the building.

3. Means of Escape

3.1 Escape Route Design
- Escape route widths and capacities determined using risk profile.
- Maximum permissible **travel distances** based on risk profile: **[Insert Distances]**.
- Escape route stair widths and final exits designed in accordance with **BS 9999 Table 2**.

3.2 Staircase Design and Evacuation Strategy
- Number and width of stairs based on occupancy load.
- Protected stairwells provided in accordance with risk assessment.
- Evacuation lift provisions for mobility-impaired occupants.

3.3 Refuge Areas
- Refuge spaces provided at **[Insert Locations]**.
- Emergency communication points provided for assisted evacuation.

4. Fire Compartmentation and Passive Fire Protection

4.1 Compartmentation Strategy
- Fire compartments designed based on risk profile and fire resistance requirements.
- **Fire resistance ratings**: [Insert Fire Rating] based on **BS 9999 Section 12**.

4.2 Fire Doors and Fire Stopping
- **Fire doors** to be installed as per **BS 9999 Table 6**.
- **Cavity barriers** and fire stopping measures to prevent unseen fire spread.

5. Fire Detection and Alarm Systems

- Fire detection and alarm system designed in accordance with **BS 5839-1 (Commercial) or BS 5839-6 (Residential)**.
- **System Category**: [Insert Type, e.g., L1, L2, L3, P1, P2].
- Smoke detection in escape routes and high-risk areas.

6. Fire Suppression Systems

- **Sprinkler system** installed as per **BS EN 12845 / BS 9251**.
- Alternative suppression systems (water mist, gas suppression) assessed as per risk profile.
- Fire-fighting facilities (hose reels, extinguishers) provided in accordance with **BS 5306-8**.

7. Smoke Control and Ventilation

- **Corridor smoke ventilation**: [Natural / Mechanical] per **BS 9999 Section 6**.
- **Stairwell pressurisation system** designed to prevent smoke ingress.
- Smoke shafts sized in accordance with **BS 7346-8**.

8. Firefighting Access and Facilities

8.1 Fire Service Access

- Fire appliance access provided per **BS 9999 Table 16**.
- Minimum **3.7m road width** for fire appliances.
- Fire service vehicle access to within **45m** of entrance.

8.2 Fire Hydrants and Water Supply

- Hydrant provision in accordance with **BS 9990**.
- Fire mains, risers, and firefighting shafts per **BS 9999 Section 21**.

9. Fire Safety Management

9.1 Fire Safety Plan

Responsible person(s) assigned under **Regulatory Reform (Fire Safety) Order 2005**.

Regular **fire drills and staff training** conducted.

Maintenance schedule for fire safety systems per **BS 9999 Section 11**.

9.2 Fire Risk Assessment

- Fire risk assessments conducted under **BS 9997**.
- Evacuation plans and procedures reviewed periodically.

10. Conclusion

This fire strategy has been developed in line with **BS 9999** to provide a comprehensive approach to fire safety. It balances design flexibility with fire protection measures to ensure regulatory compliance and occupant safety. Any deviations from prescriptive standards are justified through fire engineering principles.

Example Zone Modelling Report using B-Risk
Zone Fire Modelling Report: B-Risk

Project Title: Fire Risk Assessment for [Building Name]
Date: [Insert Date]
Prepared By: [Your Name/Organization]

1. Introduction

This report presents a **zone fire model simulation** using **B-Risk** to evaluate fire dynamics, smoke movement, and occupant tenability conditions in **[Building Name]**. The study aims to assess fire risks and optimize fire protection strategies for improved safety and regulatory compliance.

2. Objectives

The primary objectives of this B-Risk simulation include:
- Predicting fire growth, smoke layer descent, and temperature distribution.
- Assessing tenability limits (temperature, visibility, and toxic gases) for occupant safety.
- Evaluating the impact of fire suppression systems and ventilation.
- Providing recommendations for fire safety improvements.

3. Simulation Setup
3.1 Building Characteristics
- **Structure Type:** [E.g., Commercial, Residential, Warehouse]
- **Total Floor Area:** [m²]
- **Number of Compartments:** [Number]
- **Fire Compartment Volume:** [m³]
- **Fire Origin Location:** [E.g., Storage Room, Kitchen, Office Space]

3.2 Fire Source and Growth
- **Fuel Type:** [Material, e.g., wood, plastic, hydrocarbons]
- **Heat Release Rate (HRR):** [Peak HRR in kW]
- **Fire Growth Rate:** [Fast, Medium, Slow]
- **Fire Load Density:** [MJ/m²]
- **Ignition Source:** [E.g., electrical fault, open flame]

3.3 Ventilation and Fire Protection Features
- **Door and Window Openings:** [Number, dimensions, and locations]
- **HVAC System:** [Operational or non-operational]
- **Sprinkler System:** [Activation time, discharge rate]
- **Smoke Extraction:** [Mechanical or natural ventilation]

4. Results and Analysis
4.1 Fire Growth and Heat Release Rate
- The fire reached a peak **HRR of [Value] kW** in **[Time] minutes**.
- Flames spread to adjacent materials within **[Time] minutes**.

4.2 Smoke Layer Development
Smoke layer height decreased to **[Height] m** within **[Time] minutes**.

Visibility reduced to **[Distance] m** in **[Location]** within **[Time] minutes**.

CO concentration exceeded tenability limits in **[Location]** after **[Time] minutes**.

4.3 Tenability Conditions
Temperature: The upper layer temperature exceeded **[Limit, e.g., 200°C]** within **[Time] minutes**.

Visibility: Reduced below **[10m]** in key exit routes, affecting safe egress.

Toxic Gases: CO levels reached **[PPM]**, creating hazardous conditions for occupants.

4.4 Fire Suppression and Ventilation Effects
Sprinkler Activation: Reduced fire size by **[Percentage]** and delayed smoke layer descent by **[Time] minutes**.

Mechanical Ventilation: Successfully delayed tenability threshold exceedance by **[Time] minutes**.

Fire Doors: Prevented smoke migration to **[Describe protected areas]**.

5. Conclusions and Recommendations
5.1 Key Findings
- Fire and smoke spread posed **[Describe key risk areas]**.
- Tenability limits for temperature, CO, and visibility were exceeded in **[Time]**.
- Ventilation effectiveness varied based on **[Describe conditions]**.

5.2 Recommendations

- **Fire Protection Enhancements:** Improve suppression system coverage.
- **Ventilation Optimization:** Adjust HVAC and smoke extraction for better control.
- **Evacuation Planning:** Modify exit strategies based on visibility analysis.
- **Training and Drills:** Conduct periodic fire drills to ensure readiness.

6. Appendices

- **Appendix A:** Input Parameters and Fire Scenarios
- **Appendix B:** Smoke Layer Height vs. Time Graphs
- **Appendix C:** Temperature and CO Concentration Profiles
- **Appendix D:** Compliance with Fire Safety Regulations

Example Evacuation Modelling Report
Evacuation Modelling Report: BuildingExodus

Project Title: Evacuation Analysis for [Building Name]
Date: [Insert Date]
Prepared By: [Your Name/Organization]

Introduction

This report presents an **evacuation model simulation** using **BuildingExodus** to analyse occupant movement, decision-making, and egress times in **[Building Name]** during a fire emergency. The study aims to evaluate evacuation efficiency, identify potential bottlenecks, and improve emergency planning and safety measures.

Objectives

The primary objectives of this evacuation simulation include:

- Analysing occupant movement and egress routes.
- Identifying delays and bottlenecks affecting evacuation.
- Assessing the impact of fire and smoke conditions on evacuation times.
- Providing recommendations to optimize evacuation procedures.

Simulation Setup

Building Characteristics

- **Structure Type:** [E.g., Commercial, Residential, Stadium]
- **Total Floor Area:** [m²]
- **Number of Floors:** [Number]
- **Number of Exits:** [Number]
- **Exit Widths:** [Measurements in meters]
- **Stairwell Characteristics:** [Number, dimensions, and accessibility]

Occupant Characteristics

- **Total Number of Occupants:** [Number]
- **Occupant Distribution:** [E.g., randomly assigned, by department, event seating]
- **Mobility Factors:** [Percentage of disabled or slow-moving occupants]
- **Response Time Distribution:** [Time in seconds, based on occupant behaviour]

Fire Scenario and Environmental Factors
- **Fire Location:** [E.g., storage room, kitchen, electrical room]
- **Smoke Spread Rate:** [Rate of spread, affecting visibility]
- **Heat and Toxic Gas Effects:** [CO and temperature thresholds impacting movement]
- **Alarm Activation Time:** [Time from ignition to alarm activation]

Results and Analysis

Evacuation Time and Movement Flow
- The total evacuation time was **[Time] minutes**.
- The first occupants exited within **[Time] seconds**, while the last occupants exited in **[Time] minutes**.
- The average occupant speed was **[Speed] m/s**.

Bottlenecks and Congestion Areas
- High congestion occurred at **[Location]** due to **[Cause]**.
- Delays in evacuation were observed at **[Stairwell/Exit]**, increasing total egress time by **[Time] minutes**.
- Limited exit widths caused occupant queuing in **[Location]**.

Impact of Fire and Smoke Conditions
- Smoke reduced visibility below **[Distance] meters** in **[Location]**, slowing movement.
- Heat exceeded **[Temperature]°C** in **[Location]**, causing occupants to reroute.
- Toxic gas levels reached **[PPM]**, posing a risk to delayed occupants.

Effectiveness of Fire Safety Measures
- **Sprinkler Activation:** Reduced smoke spread and visibility impairment by **[Percentage]**.
- **Emergency Lighting:** Improved wayfinding, reducing hesitation time by **[Time] seconds**.
- **Exit Signage and Alarms:** Enhanced occupant awareness, decreasing pre-movement time by **[Time] minutes**.

Conclusions and Recommendations

Key Findings
- The evacuation process was **[Efficient/Inefficient]** due to **[Key Factors]**.
- The longest delays were observed at **[Location]**, significantly impacting total evacuation time.

- Fire and smoke conditions **[Did/Did Not]** substantially affect evacuation efficiency.
- The safety margin was **[State]**.

Recommendations
- **Increase Exit Capacity:** Expand or add exits to reduce congestion.
- **Improve Signage and Wayfinding:** Enhance illuminated signage for faster occupant navigation.
- **Reduce Response Time:** Implement training programs to shorten reaction time.
- **Optimize Fire Protection Systems:** Improve smoke extraction and suppression methods.

Appendices
- **Appendix A:** Input Parameters and Occupant Distribution
- **Appendix B:** Evacuation Time vs. Exit Flow Graphs
- **Appendix C:** Heat and Smoke Contour Maps
- **Appendix D:** Comparison with Fire Safety Regulations

Example CFD Report
Computational Fluid Dynamics (CFD)

Fire Simulation Report: SmartFire

Project Title: Fire Dynamics Simulation for [Building Name]

Date: [Insert Date]

Prepared By: [Your Name/Organization]

1. Introduction

This report presents a Computational Fluid Dynamics (CFD) fire simulation using **SmartFire** to evaluate fire development, smoke propagation, and occupant safety within **[Building Name]**. The study aims to assess fire hazards, smoke movement, and thermal conditions in various scenarios to optimize fire protection measures and evacuation planning.

2. Objectives

The primary objectives of this CFD fire modelling analysis include:

- Evaluating the fire growth and smoke spread within the selected structure.
- Assessing tenability conditions (temperature, visibility, and toxicity) affecting occupants.
- Testing different fire protection strategies (e.g., ventilation, suppression systems).
- Providing recommendations for design and emergency planning.

3. Simulation Setup

3.1 Building Geometry and Layout

- **Structure:** [Type of building, e.g., multi-story office, warehouse, residential complex]
- **Total Area:** [Square meters]
- **Fire Compartment Details:** [Number of rooms, openings, ventilation features]
- **Fire Location:** [Origin of fire, e.g., electrical room, kitchen, storage area]

3.2 Fire Source Characteristics

- **Fuel Type:** [Material, e.g., wood, plastic, hydrocarbon-based fuels]

- **Heat Release Rate (HRR):** [Peak HRR, growth rate, decay phase]
- **Fire Load Density:** [MJ/m²]
- **Ignition Source:** [E.g., electrical fault, open flame]

3.3 Simulation Parameters

- **CFD Model:** Large Eddy Simulation (LES) / Reynolds-Averaged Navier-Stokes (RANS)
- **Grid Size:** [Cell resolution used for computational domain]
- **Time Step:** [Simulation duration and time increment]
- **Ventilation Conditions:** [Natural vs mechanical, exhaust system specifications]
- **Material Properties:** [Wall, floor, and ceiling materials used in fire modelling]

4. Results and Analysis

4.1 Fire Growth and Spread

- The fire reached a peak HRR of **[Value] kW** within **[Time] minutes**.
- Flame spread followed **[Describe spread pattern, e.g., through open doors, HVAC systems]**.
- The upper layer temperature exceeded **[Value]°C** within **[Time] minutes**, posing a high risk to occupants.

4.2 Smoke Propagation

- Smoke travelled through **[Describe movement through stairwells, corridors, or vents]**.
- Visibility dropped below **10m** in **[Location]** after **[Time] minutes**, impacting evacuation.
- Carbon monoxide (CO) concentration exceeded tenable limits in **[Location]** after **[Time] minutes**.

4.3 Tenability Conditions for Occupants

- **Temperature Thresholds:** Heat exposure exceeded **[Limit, e.g., 60°C]** in **[Location]**.
- **Visibility Levels:** Smoke obscured exit routes within **[Time] minutes**.
- **Toxic Gas Accumulation:** CO levels reached **[PPM]** within **[Time]**, posing fatal risks.

4.4 Effectiveness of Fire Protection Systems

- **Automatic Sprinklers:** Activation at [Time] minutes reduced fire growth rate by [Percentage].
- **Mechanical Ventilation:** Effectively delayed smoke descent by [Time] minutes.
- **Fire Doors:** Prevented smoke migration to [Describe protected areas].

5. Conclusions and Recommendations

5.1 Key Findings

- Fire and smoke spread posed [Describe main risk areas].
- Tenability thresholds for temperature and toxicity were exceeded within [Time].
- Ventilation effectiveness depended on [Describe impact of natural vs. mechanical systems].

5.2 Recommendations

- **Enhanced Fire Protection:** Install additional suppression measures (e.g., sprinklers, fire-resistant barriers).
- **Optimized Ventilation:** Implement mechanical exhaust systems to improve smoke control.
- **Evacuation Strategy Improvements:** Improve wayfinding, emergency lighting, and signage.
- **Training and Drills:** Conduct regular evacuation drills to prepare occupants for emergency scenarios.

6. Appendices

- **Appendix A:** Simulation Parameters and Input Data
- **Appendix B:** 3D Visualizations and Temperature Contour Maps
- **Appendix C:** Tenability Graphs for Smoke, CO, and Temperature
- **Appendix D:** Comparison with Fire Safety Regulations

Fire Investigation Report

Prepared in accordance with UK legislation and best practices

1. Report Details

- **Report Number:** [Insert Number]
- **Date of Report:** [DD/MM/YYYY]
- **Investigating Officer(s):** [Name(s) and Title(s)]
- **Authority/Agency:** [Fire and Rescue Service, Private Investigator, etc.]
- **Incident Number:** [If applicable]
- **Date and Time of Incident:** [DD/MM/YYYY, HH:MM]
- **Location of Fire:** [Full Address]

2. Executive Summary

A brief overview of the incident, including key findings and conclusions.

3. Incident Details

- **Type of Premises:** [Residential, Commercial, Industrial, etc.]
- **Occupancy at the Time of Fire:** [Number of People Present, Usage]
- **Reported Cause:** [If known]
- **Weather Conditions:** [Relevant Weather Data]
- **Emergency Services Involvement:** [Fire Service, Police, Ambulance, etc.]

4. Fire Scene Examination

- **Initial Observations:** [State of the Scene, Damage Overview]
- **Fire Patterns and Spread:** [Fire Growth, Smoke Patterns]
- **Structural Damage Assessment:** [Damage to Building and Safety Risks]
- **Possible Ignition Sources:** [Electrical, Open Flame, Arson, etc.]
- **Presence of Accelerants:** [If Applicable]
- **Fire Safety Measures:** [Smoke Alarms, Sprinklers, Fire Doors]

5. Witness Statements

- **Witness 1:** [Name, Statement Summary]
- **Witness 2:** [Name, Statement Summary]
- **Other Relevant Witnesses:** [Details]

6. Fire Investigation Analysis
- **Cause and Origin Analysis:** [How and Where the Fire Started]
- **Possible Contributing Factors:** [Faulty Wiring, Human Error, etc.]
- **Compliance with Fire Regulations:** [Regulatory Adherence]
- **Fire Load and Fuel Sources:** [Materials Contributing to the Fire]
- **Timeline of Events:** [Sequence of Fire Development]

7. Findings and Conclusion
- **Probable Cause:** [Identified Cause]
- **Intentional or Accidental:** [Arson, Accidental, Natural Causes]
- **Regulatory Breaches Identified:** [Fire Safety Failures, If Any]
- **Recommendations for Prevention:** [Changes Needed, Safety Enhancements]

8. Supporting Documentation
- **Photographs:** [Attach as Appendix]
- **Fire and Rescue Service Reports:** [Attach Copies]
- **CCTV Footage (If Available):** [Summary of Footage]
- **Forensic Evidence:** [Laboratory Results, If Applicable]
- **Previous Fire Safety Inspections:** [Compliance History]

9. Legal Considerations
- **Relevant UK Legislation:**
- Regulatory Reform (Fire Safety) Order 2005
- Fire and Rescue Services Act 2004
- Health and Safety at Work Act 1974
- Building Regulations (Part B: Fire Safety)
- BS 9999: Code of Practice for Fire Safety in the Design, Management & Use of Buildings
- **Potential Legal Proceedings:** [Prosecution, Civil Liability, Recommendations]

10. Report Approval
- **Lead Investigator:** [Name, Signature, Date]
- **Reviewing Officer:** [Name, Signature, Date]
- **Agency/Organisation Approval:** [Official Stamp/Approval, If Required]